Funding Windows and Research In India

Funding Windows and Research in India

Sharat Kumar
Minnie Mathew

WFP

United Nations
World Food Programme

SHIPRA

SHIPRA

ISBN : 81-7541-220-8

First Published in India in 2005

Funding Windows and Research In India

Published by :
SHIPRA PUBLICATIONS
115-A, Vikas Marg, Shakarpur,
Delhi-110092 (India)
Ph. : 91-11-55277210, 22500954
Email : info@shiprapublications.com
www.shiprapublications.com

Laser Typeset by :
Mrigakshi Designers
Delhi-110092

Printed at :
Chaudhary Offset Process,
Delhi-110051

Preface

The Government of India lays great stress on the development of social sector. The Common Minimum Programme which was released on May 2004, mentions that academic excellence and professional competence will be the sole criteria for all appointments to bodies like the Indian Council for Historical Research, the Indian Council for Social Science Research, the University Grants Commission, the National Council for Educational Research and Training etc.

I am impressed to note that there is an inbuilt provision for research in the various development sectors in India. This reflects the importance ascribed by the Government of India to attain excellence in all spheres of its development initiatives. Although some Multinational Companies are setting up their research centers, it will take some time before they move their cutting-edge research facilities to India. As such, the Government offers the best opportunity for research for most Indians. However, while great research opportunities exist in India there are limitations in accessing them.

There is a great demand from students to know about the various research opportunities. The effort of the authors in presenting this wealth of information in a user-friendly manner in a single book is indeed commendable. I would say this publication is a gift to every student of research. The book offers an entry point to information and resources related to research, both for academics and practitioners.

India has the largest food assistance programmes in the world and I see a great potential for research in this area for understanding the factors that influence successful programme implementation. There are several success stories to be shared with the rest of the world. From the perspective of the World Food Programme, whose mandate is food based programmes, research findings on this aspect is very important. I think India has a lot to offer in terms of development research.

What began as a simple survey of various funding windows in India for food assistance programmes, has turned out to be a comprehensive publication which provides an anthology of funding windows for research covering all sectors of the economy encompassing both social and physical sciences in India. It also provides an overview of the policy for research promotion and examines the future of research in India.

Pedro Medrano

Acknowledgement

This study owes the most to Mr. Pedro Medrano, Country Director of the World Food Programme (India), *at whose behest the work was initiated.* Food security is at the heart of the development agenda of the UN World Food Programme. The study was initiated for surveying the various 'funding windows' available in India for doing research *for strengthening the effectiveness of food assistance programmes.* It is worth mentioning, in this regard, that India has one of the biggest food assistance programmes in the world and there always has existed a feeling that the returns are not commensurate to the expenditure incurred.

This short study vis-à-vis the survey was soon completed and submitted to the WFP. It was, however, felt the study could be developed to take stock of all the 'funding windows' in India supporting *research.* With encouragement from the UN World Food Programme and the necessary permission granted by the Planning Commission/ Government of India, the original project could be enlarged and given the shape of a book. The findings of the study are based on various reports (published and unpublished), books and articles, which have been enlisted in the Select Bibliography.

While Chapters II, III, and IV owe greatly to Dr. Sharat Kumar, Chapters I, V and VII are the result of the combined effort of Dr. Minnie Mathew and Dr. Sharat Kumar. Besides them, the book has been benefited from individual and combined efforts of several people. Prof. Hans Kumar Sahai provided the input relating to the field level perspective, in accessing the different 'funding windows'. Ms. Shalini Sanan and Mr. Vineet Prasad as research associates, did the arduous task of collecting and tabulating the data for Chapter V and also compiling the essential features of the various 'funding windows' within India and abroad. Chapter VI on 'Future of Research' is the work of Dr. Ashish Kumar.

The book owes a great deal to the followings for making available relevant documents and for their enthusiastic support through discussions, namely, Mr. Wolfgang Herbinger, Mr. Shankar Prasad, Mr. J. Srivastava, Dr. Ramesh Sharan, Mr. P. V. R. Sebastian, Mr. M. R. Anand, Ms. Veena Prasad, Mr. S. K. Roy, Prof. K. K. Sinha, Mr. B. Gangaiah, Mr. Shishir Kumar, Prof. R. C. Sinha, Mr. Wasim Akhtar, Dr. Paul Kattuman, Ms. Radha R. Ashrit, Dr. P. C. Verma, Mr. V. K. Nayyar, Dr. Nagesh Singh, Mr. Srikanta K. Panigrahi and Ms. Kirti Saxena. We acknowledge our sincere gratitude to Prof. Rajeev Verma for going through the text for editorial corrections and for other valuable suggestions.

The *'funding window'* for the present study was none other than the World Food Programme (United Nations); the views and opinions expressed are, however, those of the authors alone. The staff of the various libraries like those of the Planning Commission, the Ministry of Finance and the National Institute of Public Finance provided their best cooperation and we are indebted to them. We owe special thanks to Mr. J. L. Kishnani for typing the manuscript. Ms. Rakhi P. Bharadwaj has done the cover design of the book.

Contents

List of Tables

List of Grants-in-aid of Government of India

List of Fellowships available to Indian Students and Scholars

Chapter 1

Introduction

Research and Development (R&D), in post-independent India, can be divided into two phases. The first phase was characterized by the criticality of 'self-sufficiency' in the field of technology development for rapid industrial and agricultural development. The second or the current phase is characterized by the desire to take advantage of 'globalization'. The second phase is the outcome of the post-1990 'economic liberalization', and is characterized by progressive reduction in tariffs both *for imports and exports*. Possibilities of higher exports exist, but these can materialize if the competitiveness of (Indian) goods and services is considerably enhanced. Competitiveness, in turn, would depend greatly on investments in research. While the first phase witnessed the policy announcements such as the Scientific Policy Resolution (1958) and the Technology Policy Statement (1983) and a pro-active role of the Government in setting up various institutions of scientific research, the second phase is characterized by *interest in research* by both the public and the private sectors.

More than fifteen thousand crore of rupees are presently spent annually on research and development (R&D) in India. Three-quarters of this expenditure, so far, comes from the government. It is important to note that *the share of the private sector, which stood at less than 5 per cent during 1966, went up to nearly 25 per cent of the total expenditure on R&D by 1998. In specific sectors, such as, biotechnology, pharmaceuticals, telecommunication and information technology, private sector investments in research are more than the public sector investment.* Altogether, this works out to around 1-2 per cent of the national income of India. Some of the other developing countries like Brazil, China, Israel, Korea, Singapore and Taiwan, spend more in percentage terms than India on research and development. Developed countries, in general, spend more than the developing countries on R&D. It is worth recognizing that while producing an item is often within everyone's reach, *developing a product* is expensive and prohibitive as it involves considerable research and expenditure.

Although considerable differences exist between Karl Marx and Joseph Schumpeter in their respective treatments of 'capitalism', both consider *technical innovation* an essential feature of the capitalist system. Both recognize that capitalism should be seen as a dynamic process of 'creative destruction', *where the old were continuously being replaced by the new*. According to Schumpeter, 'the real competition for a small business does not emanate from more competitors per se; *it is the new product, new way of doing things, new technology or type of organization*' (Ryon, 1994). The economist N. Kontratieff, similarly, noted fifty-year waves of *continuous economic expansion under capitalism*, which he attributed to clustering of innovations. Thus, innovations of iron smelting by pit-coal and mechanization in the cotton industry were responsible for the first wave and the innovation of steam power, steel and the railways for the second wave. Together, they engineered what is known as the 'Industrial Revolution'. Subsequently, innovations in electric, chemical and automobile sectors led to the third wave and innovations in electronics, synthetic materials and petrochemicals resulted in the fourth wave of economic expansion. 'Some even predict a fifth long wave based on information technology' (Ryon, 1994).

Countries willing to catch up with those ahead of them have been assigning a great deal more significance to 'research'. As it has been observed, 'in the sixties, the Japanese were not technological leaders. The country had to import technologies in a major way. But the Japanese made it a point, *mostly through voluntary action by their industries and government agencies*, to invest about four times more towards their own technology development for every unit of money they spent on importing technology' (Kalam, 1998). India's share of export in world trade is around 0.7 per cent, while that of China is 4 per cent. India's low share in world trade is observed to be on account of the low magnitude *of high technology exports*. The Steering Committee Report on Science and Technology (Tenth Five Year Plan), Planning Commission, similarly observes, 'Fortunately today the Indian industries and the science and technology sectors are favourably inclined towards each other. Industry should take advantage of this situation and by maintaining perpetual interface with academia continuously upgrade its technologies, quickly responding to market demands'. According to one estimate, India would need a workforce of around 50,000 in the field of science and technology (S&T) alone during the Tenth Five Year Plan (2002-2007).

Besides trade and commerce, 'research' relating to socio-economic aspects helps increase awareness about social moorings. It also helps in identification of problems and solution to problems and in identifying the institutional reforms necessary for a better future. Institutions of higher learning have been known to play a crucial role in national progress, self-reliance and security as well as in the cultural and economic progress of the society. Universities constitute the most important part of academia. The universities in India, however, have not had adequate funds *of their own* for financing researches in different disciplines. In the opinion of the 'National Policy on Education (NPE) 1986', however, 'Teaching and research should go hand-in-hand in higher educational institutions. *It is important that teachers get an opportunity to involve themselves in research activities.* In order to do so, there should be a separate budget for research within the budget of educational institutions'. It also recommended constituting a Board or having a Dean to oversee the disbursement of grants-in-aid for research. The Tenth Plan also observed, 'policies, procedures and systems should be reformed to encourage the academic faculty to accept contracts or do collaborative research for industry'.

The University Grants Commission (UGC) has been the main funding agency for research in the universities. The UGC, which was established for coordination, determination and maintenance of standards in university education, has been providing financial assistance to all eligible central, state and deemed universities for improving their infrastructure and basic facilities for research. It also provides grants to *university departments* under its Special Assistance Programme (SAP) for research and quality teaching. Financial support is provided depending upon the level (DRS, DSA, CAS) at which the departments have been placed by the Expert Committee constituted by the UGC. Under its scheme of Assistance for Strengthening of Infrastructure in Science and Technology (ASIST), the UGC has been providing financial support to select departments in science and technology to acquire sophisticated and expensive equipments. The UGC has also set up Inter-University Centres to provide state-of-the-art equipments and library facilities. The Commission has in recent years also embarked upon the concept of *'institutes of excellence'* within the university system *for research studies*. It has, similarly, been providing financial grants to universities, postgraduate colleges as well as non-university institutions for organizing national and international workshops, symposia, seminars and conferences.

The UGC has been conducting the national entrance test (NET) since1984 for awarding Junior Research Fellowships (JRFs) to young researchers to pursue research leading to M.Phil. and Ph.D. degrees. The number of research fellowships funded by the UGC is to be further increased during the Tenth Plan. (The drop out rate among the research fellows qualifying for NET is, however, fairly high at 35 per cent.). Selection tests for JRF in science subjects are conducted in collaboration with the Council for Scientific and Industrial Research (CSIR). Under the scheme of Research Awards, the Commission enables the teachers who are doctorates and working on a permanent basis to pursue research in their area of specialization without having to undertake teaching responsibilities. The Memorandum of Understanding (MoU) signed between the UGC and the CSIR in 1991 permits scientists in universities to access the research facilities of CSIR. According to the Tenth Plan, moreover, 'policies' procedures and systems should be reformed to encourage the academic faculty to accept contract for collaborative research for industry'.

Apart from the UGC, *the various research councils* have been contributing significantly to promotion of research. These are autonomous organizations set up by the Government of India. Prominent research councils in the social sciences are the Indian Council of Social Science Research (ICSSR), the Indian Council of Historical Research (ICHR), the Indian Council of Philosophical Research (ICPR) and the Indian Institute of Advanced Studies (IIAS). The Indian Council of Agricultural Research (ICAR), the Indian Council of Forestry Research (ICFR) and the Indian Council of Medical Research (ICMR) are the other apex bodies for coordinating research in areas relating to agriculture, forestry and bio-medical and health sectors respectively. The ICAR works in close collaboration with the state agricultural universities and central (agricultural) institutes in India. Teaching, research and extension are the three primary objectives of the agricultural universities. The ICMR on the other hand has been conducting in-house research through institutes under it, which are located in different parts of the country. These institutes specialize in areas of communicable diseases, non-communicable diseases, maternity, nutrition, child health, traditional medicine and basic research. It also provides grants for senior research fellowships and research associateships to outside scholars for doing research either at the permanent institutes, medical colleges, universities or other biomedical research institutes in India.

Besides setting up a network of in-house research institutes in different areas, the Council for Scientific and Industrial Research (CSIR) has been providing support for sponsored extramural research in universities and other institutes of higher learning. *It also provides fellowships and associateships to young researchers.* The CSIR, moreover, has research establishments (institutes and laboratories) directly under its umbrella that are located in different parts of the country. These establishments specialize in areas as diverse as 'aerospace' to 'biotechnology'. Biotechnology or *application of knowledge of biology in solutions of problems in agriculture and medicine* is believed to have the potential to revolutionize human welfare through improvements in food, health and environment. The National Centre for Plant-Genome Research and the National Brain Research Centre are the other autonomous institutes recently established by the Department of Bio-technology, Government of India.

Research is also undertaken in the other centres of higher education, namely, the Indian Institutes of Technology (IITs) and the Indian Institutes of Management (IIMs). Under the Institutes of Technology Act 1961, IITs have been set up in the different parts of the country (originally, at Delhi, Kanpur, Kharagpur, Madras and Mumbai) as institutions of national importance. The main objective of the IITs is to impart world-class training in engineering and technology. They also conduct research with extramural support from CSIR and consultancies to the public and private sector enterprises. As part of the research activity, IITs have been offering M.Phil., M.Tech. and Ph.D degrees on lines similar to those of the universities. Likewise, the Indian Institutes of Management (IIMs) located at Ahmedabad, Kolkata, Bangalore, Lucknow, Indore and Kozhikode also conduct research and provide consultancy services in the field of management apart from their main activity of teaching. The Indian

Institute of Science (Bangalore), the National Council of Applied Economic Research (New Delhi), the Institute of Economic Growth (Delhi), the National Institute of Rural Development (Hyderabad) and numerous other specialized institutions at the national and state levels have been conducting *in-house* research. Major research centres were established during the Ninth Plan period at select centres of excellence, namely, the National Centre for Computational Fluid Dynamics at IIT, Chennai, the Technical Accoustics facilities and x-ray facility for Structural Biology at IIT, Bangalore and Laser Scanning Confocal Microscope Facility at BHU, Varanasi.

The Planning Commission has been providing financial assistance for research studies *to institutions and scholars* for conducting research from the very beginning of planning in India. The various Ministries/ Departments under the Government of India, namely, the Department of Atomic Energy, Space, Electronics, Science & Technology, Ocean Development and Environment & Forest on the other hand are composed of a cadre of scientists, and research is conducted (mainly) *in-house*. The Department of Ocean Development has been conducting multi-disciplinary research *on board the two research ships Sagar Kanya and Sagar Sampada* (to be replaced now by new ships) to study the physical, chemical, geological and biological aspects of the Indian Ocean. The Vikram Sarabhai Space Centre under the Department of Space has been making contributions towards the indigenous development of rockets and satellite launcher vehicles with their associated control and guidance systems. The National Remote Sensing Agency (Hyderabad) has been responsible for acquisition, processing and dissemination of satellite and aircraft remote sensing data. The Indian National Satellite System (INSAT) under the Department of Space has been providing space services in the areas of telecommunication, TV broadcasting and meteorology. The earth observation system is similarly oriented towards building state-of-the-art satellite systems (configured for application related to land and ocean), *to address national needs as well as global service requirements.* The Bhaba Atomic Research Centre and the Tata Institute of Fundamental Research under the Department of Atomic Energy have been contributing towards development of control and instrumentation systems for nuclear power plants being set up by the Nuclear Power Corporation of India (NPCIL). The Defence Research and Development Organization (DRDO) has been conducting research in areas of aeronautics and missiles in India. Altogether there are more than two hundred national laboratories and more than one thousand R&D bodies in the industrial sector.

Some initiatives towards *international cooperation* in research and development have been taken in the direction of negotiating foreign foundations offerings (fellowships) in science and technology that may be centrally operated. Institutions like the Indian Institute of Public Administration (New Delhi) have gone for strategic partnership with research institutes abroad in Malaysia and Mauritius. 'At the initiative of the United Nations, India has established a Centre for Space Science and Technology Education for Asia and the Pacific (CSSTE-AP) at Dehradun' (Planning Commission, SCSTTP, 2001). Similarly, while the Welcome Trust Grant type of fellowships is proposed to be offered to Non-resident Indians (NRIs) and persons of Indian Origin (PIOs), *initiatives are being taken to throw open the institutions of excellence to foreign nationals* for utilizing facilities, such as, those of the Giant Meterwave Radio Telescope (SMRT) at Pune and the Telscope Facility at Hanley (Ladakh).

Multilateral and bilateral agencies like the World Bank, UNICEF, WHO, WFP, UNDP, European Union, JBIC, British Council, USAID, SIDA, DANIDA, CIDA and NORAD etc have been making their contribution to research in India through organizing seminars and training programmes for transfer of technology and capacity building in different sectors. Next to the multilateral and bilateral donor agencies are the NGOs, both national and international, who have been contributing to research in India. While international NGOs such as the Ford Foundation and the Rockefeller Foundation have been awarding fellowships to research scholars on a regular basis *in areas of their interest*, the national NGOs such as the MS Swaminathan Research Foundation and various other NGOs and Trusts such as the PUDR, PUCL, SEWA and PRIYA have been adding to the stock of knowledge with their in-house research findings and investigative reports.

Research may be pursued as a result of personal preference, on account of institutional policy or for the gains expected from the market. It may be broadly in areas of basic research or applied research. Very often the research findings of research may remain unnoticed.

This may happen especially in the case of work by young researchers where the subject is based on individual choice, *rather than on demand from the industry or the government*. The intrinsic value of research in these cases lies more in acquiring skills in the research methodology, and the young researchers contribute to the society in the long run. Nevertheless, as argued by the National Policy on Education, 'Research of an indifferent quality neither contributes to knowledge nor to any social good; such research will have to be guarded against and reduced to the minimum'. It has been also pointed out that adequate attention was not paid to 'intellectual property rights' (IPR) in the various policy declarations by the Government until recently and this contributed to the lack of enthusiastic response to the pursuit of research in India. For the first time, there is a clear recognition of IPR in the Science and Technology Policy announced in 2003.

Dissemination and utilization of the findings of research is equally important. Seminars, training programmes and publications contribute eminently towards this objective. Libraries, presence of an active media (newspapers and T.V), universal primary/ general education, all serve this objective in their own unique ways. Libraries have a special role in research since they not only help transfer knowledge but help the researcher in reviewing the existing stock of knowledge *at the start of his work*. In a country like India, moreover, which has a large illiterate population *it is the village extension worker* or *the angadwadi worker* who become the torchbearers for the ignorant farmer or the uninitiated mother *in imparting the best practice.*

Objectives and Structure of the Book

The present study has in the main the following objectives:

(a) to provide an overview of the various funding windows for *extramural research (other than in-house research)*

(b) to review the general policy for research promotion and dissemination and

(c) to examine the future prospect of research in India.

Chapter V provides an anthology of the major funding windows of the Government of India. It also provides information on the expenditure on research by each of the funding agencies during the past five years. Chapter V also includes a case study of one of the universities to understand the field level dynamics in accessing the different funding windows. Chapter III and IV of the book dwell upon the *government policy* towards promotion and dissemination of research. Development of research in post-independent India has similarly been discussed in Chapter I. Chapter II provides briefly the essence and principles of research. It discusses the approach to research, its relationships to knowledge, research as a means to discovery and its limitations. Chapter VI discusses the future of research in India and examines the prospects of India emerging as an important global centre for research.

Chapter 2

Basics of Research

Research supposedly contributes to the well being of men and women through (a) greater awareness of physical and social realities (b) introduction of new products and services and (c) higher productivity/ better technology. Research may also be pursued for the sake of enquiry. The philosophers of ancient Greece considered even disinterested enquiry as ethically good.[1] It has further been observed, 'A hierarchy which lives by dogmatic system of belief may to a large extent speak with one voice on all manner of things when enquirers held various opinions' (Russell, 1959).

The renaissance movement in Europe during the 15-17th century too had, in good measure, its origin in the outlook of the Greeks. In this tradition there was no difference between the work of the artist and the scientific enquirer. Both in their several ways sought after truth whose essence was grasped through numbers (Russell, 1959). The great out burst of creativity during the renaissance thus owed greatly to mathematics. The underlying principle of proportions in human body, in flowers in nature, in music, all seemed to possess an underlying numerical relation. If further explored this could provide the key to the design of the universe. This spirit of renaissance inspired great paintings, sculptures and architecture. The "Principle" of Newton (1642-1727) also proceeded in the manner of "Elements" of Euclid (1300 B.C.). 'By purely mathematical deduction from the three laws of motion and the laws of gravitation, it explained the whole solar system' (Russell, 1959).

In the Indian context, the following observations, in regard to mathematics are worthwhile mentioning namely 'Medieval Indian mathematicians, such as, Brahmagupta (7th century), Mahavira (9th century) and Bhaskar (12th century), made several discoveries which in Europe were not known until the Renaissance or later' (Basham, 2003). The Sulbe Sutras (rule of the thread of measuring chord) which form part of Kalpasutras amongst the Vedangas contain rules regarding enumeration, arithmetical operations, fractions, properties of rectilinear figure — *the so called Pythegorean theorem*, irrational numbers (quadratic and indeterminate). 'The canonical literature of the Jains also laid great emphasis on mathematics and enumerates various topics such as number reckoning, fundamental operations, geometry, mensurations, fractions, equations, permutations and combination' (Kuchhal, 1995).

2.1 Knowledge: Sensible and Intelligible

Knowledge, according to Socrates (470-399 B.C.) is true judgment, *fixed and certain*, a truth free from error. All particular things (objects or events) that we grasp through our *senses* are, however, in continuous flux in time and space. '*One cannot step twice into the same river, for fresh waters are ever flowing in' (*Heraclitus, Sixth Century B.C.)[2]. What is 'sensible' is, therefore, transient and apparent. Plato *(428-348 B.C.)*, nevertheless, provided a solution to *the puzzle that everything is in a flux* through bringing about a synthesis between 'motion' and 'rest' into *"being"*. Being subsumes into itself (both) *"what is, is not"*.

All *judgments about particulars* thus fail to qualify as *knowledge* as defined above *and* can best be called "opinions". Knowledge gained from mathematics is, however, independent of sense perception. The derivations proceed from self-evident premises and there are no facts outside itself that call for comparisons. Moreover, 'when a mathematician proves a particular proposition about triangles, rather it is something he sees in *mind's eye' (Russell, 1959)*. Quite similarly *ideas* are objects imprinted on the empty-space, that is, *irrespective of time and space.* In comparison to our perception of particular things, ideas refer to the 'ideal' or the 'universal form' of the particular. 'Thus the general word "horse" refers not to this horse or that horse, but to any horse. It's meaning is independent of particular horses and what happens to them, it is not in space and time, but eternal' (Russell, 1959).

The concept of ideas as developed by Socrates belongs to this category of what is 'intelligible'. Pythagoras, moreover, argued that it is the units what gives measure to the (boundless) world. 'Numbers were thought of as made up of units, and the units, represented by dots, were taken as having spatial dimensions' (Russell, 1959). The theory of ideas with its origin in Pythagorean mathematics was, however, subjected to severe criticism by the Eleatics*. Adopting the method of deduction, Zeno of Elea (490 B.C.) showed that one reached two conclusions *about the world* based on the principle of units/ numbers. As Zeno argued, *if it were so that things are many*, then they will have to be small and large at the same time. 'Indeed, they must be so small as to be *without size* for finite divisibility shows that the number parts is infinite, *which requires units without magnitude*. But at the same time, the units must have some magnitude, and therefore things are *infinitely great*' (Russell, 1959).

Insurmountable problems arise, comparing the 'particular' with the 'universal'. Moreover, if the forms are by themselves, unrelated to our world, they are vacuous; if on the other hand they are so related, then they cannot belong to a world of their own' (Russell, 1959). Attention, in a way, reverts to seeking *"knowledge" in terms of giving an account of the sensible world around.* This is akin to the 'positive philosophy' (positivism) of Auguste Comte (1798-1857) *who insisted that one should begin with what is related to experience and need not go behind the phenomena.*

2.2 Propositions and Concepts

While *sense experience* help us acquire 'perception' about a thing, mind plays its own role interpreting the perception through *reasoning.* Any sense experience is, moreover, contingent to time and space. Since time and space exist prior to sense experience, these are considered *'a priori'* concepts. Perceptions derived from sense experience, on the other hand, are called *'a posteriori'* concepts.

The other 'a priori' concepts, which the human mind supplies to draw inferences about an object/event have been variously classified by philosophers from Aristotle (384-323 B.C.) to Emanuel Kant (1724-1804) amongst the Western philosophers into broad categories such as relation (cause/effect), quantity (unity/plurality) quality (affirmative/ negative) and modality (possibility/impossibility).

Knowledge is, moreover, generally propositional in character, being a statement having a subject and a predicate. A proposition is

* Eleatics implies Greek philosophers hailing from Elea on the west coast of present Italy.

said to be 'analytic' if the predicate is contained in the subject, for example, 'the ball is round'. It is said to be 'synthetic' when subject and predicate are distinct, for example, 'the ball is yellow'. While in the first example, a ball cannot be understood without it being round in shape (and stating that the ball is round is tautological), in the second example, the predicate mentions a particular quality/colour about the subject.

Moreover, according to Kant, while 'a posteriori' perception may be co-extensive with a synthetic proposition, 'a priori' concept is co-extensive with an analytic proposition; and they do not cut across each other. However, mathematical propositions are 'a priori synthetic' for example, 5+7=12. This proposition is 'a priori' since it is not based on/derived from sense experience. It is 'synthetic because the concept 12 (predicate) is not contained in 5,7 and + (subjects).

2.2.1 Means of Right Cognition

According to the Nyaya system of (Indian) philosophy, perception *(pratyaksha)* inference (*anumana*) inference by analogy (*upman*) and pronouncement of a reliable authority (*sabda*) *are the four means of gaining reliable knowledge* (*pramana*). The study of process of inference led to the emergence of schools of logic. False arguments were classified as *reductio ad absurdum* (*arthaprasanga*), *circular argument* (*cakra*), *infinite regression* (*anavastha), dilemma* (anyonyasraya) and *ignoretio elenchi* (*atmasraya*) (Radhakrishnan, 1957). The correct inference, under the Nyaya system is based on *"pencavyava"* form of syllogism. The five constituents of this sylloligism being, proposition *(pratijna),* reason (*hetu),* example (*udharana*), application (*upanaya*) and conclusion (*nigamana*) (Basham, 2003).

2.3 Logic and Research

Logic helps construct arguments to prove or disprove a proposition. It affords the criteria for discrimination and demonstration and is integral to any enquiry. Inferences, in turn, may be drawn based either on *deduction*, which is same as explanation or based on *induction*, which is based on observation. 'From observations of particular facts, it arrives by induction *at a general law*, and by deduction from the general law *other particular facts are inferred' (Russell, 1959).* We have thus the two broad classifications of 'deductive logic' and 'inductive logic', as it obtains in the philosophy of the West.

2.3.1 Deductive Logic

Deductive logic or method in deduction comprises starting from one or more propositions (also called premises) and then deducing other propositions in steps that follow from it to reach a conclusion. 'Syllogism' has often been considered as the most fundamental type of deductive logic. It is an argument with two *subject predicate* propositions that have one term in common and the middle term disappears in the conclusion. *For example*, if we have the first proposition as 'all men are mortal' followed by a second proposition, as 'Altaf is a man'; the conclusion reached is 'Altaf is mortal'. The middle term men/man disappears in the final argument. "Syllogism" kind of argument could likewise be made up of *modal propositions* containing 'may be' or 'must' instead of 'is' as in the above example. Bertrand Russell (1872-1970) further refined deductive logic to include relational arguments important in mathematics, such as if A>B and B>C, then A>C.

'Syllogism' as a form of argument has, however, been criticized on the ground that to know the first proposition, one would already have to know the conclusion, so that the argument begs the question. In other words, there is an inherent circularity in argument. Moreover, in the example above, we begin from a *universal proposition*, namely, 'all men are mortals', and then proceed to a *particular proposition* of 'Altaf is mortal'. According to Russell, *one need not begin from a universal proposition for reaching a correct judgment about the particular*. It is often enough to look at a single example to see the connection. For the axiom all triangles have the sum of their angles equal to two right angles', *it is enough to provide the proof with the help of a single example*.

It has been further argued that while the conclusions may follow (may be deduced) from the original premise or premises, one may reach wrong conclusions simply because the premise/premises were themselves inadequate/wrong. For example, to have an argument such as, no mammal can fly, all pigs are mammals, therefore, no pig can fly, is a valid first figure argument. While in this particular example the conclusion is true, one of the premises is false since "bats" are mammals and yet can fly (Russell, 1959). More importantly, syllogism kind of argument leaves the premises unproved.

2.3.2 Inductive Logic

Inductive logic draws inferences from observations. Induction may thus proceed from specific *(particular)* to general conclusion/ generalization. According to Francis Bacon (1561-1620), *through the process of enumeration and elimination*, one could reach significant conclusions about the object/issue under investigation. He, therefore, recommended for drawing up (a) list of things that shared a given quality (b) list of things that lacked this quality and (c) list of things that possess this quality in different degrees. In this way, it was thought one could discover the peculiar character or a quality.

Inductive Inference through experimental enquiry

To further refine the logical structure of elimination and enumerative induction, John Stuart Mill (1806-1873) elaborated a set of rules that he believed formed the basis of discovery. *Experimental enquiry is based on observation under controlled conditions.* It may thus involve various trials to confirm or disprove something, which is not explicit or is doubtful. 'Mill differentiated among the method of agreement, the method of difference, the method of residues and the method of concomitant variations. The first two, have become the basis of modern *experimental designs,* with emphasis upon comparison of controls and experimental groups. The method of concomitant variations finds contemporary expression in the correlation techniques of statistics' (Sjoberg, 1968).

2.4 Research and Hypothesis

Research is a systematic investigation and broadly consists of steps like defining the problem, formulating a hypothesis and finally reaching conclusions. R*esearch* may be explanatory, exploratory or descriptive. *Explanatory research* attempts to provide explanation of things in the form of causal relationship. *Exploratory research*, on the other hand, seeks to gain new insights or to assess a phenomenon in a new light. As against these two, *the descriptive research* involves extensive portrayal of the object/issue under consideration and is considered more appropriate for researchers who have an extensive previous knowledge of the subject. In actual practice, however, there is always a mixing and matching of the different purposes, which is also referred to as '*triangulation*'.

Setting out the Hypothesis

Research may thus be guided by the need to validate the truthfulness of the hypothesis. The *hypothesis to account for a phenomenon* may be formulated on the generally accepted view, on existing information or a mere supposition. If the consequences that follow from the hypothesis do justice to facts, the hypothesis becomes a valid explanation. If, however, the facts do not agree with the consequences following from the premises, the hypothesis gets rejected/destroyed, and paves way for another hypothesis.

Setting out the hypothesis is particularly essential to explanatory research. The hypothesis becomes the starting point for explaining a phenomenon. In regard to *exploratory and descriptive research*, the need for hypotheses arises for deciding the categories and the collection of data. Formulating a hypothesis is, however, not *sine qua non* to research. As informed, 'almost no contributor to this volume (of social science research) has been guided by hypothesis, and some confess that their theorizing was only "post factum" ' (Bose, 1995). The theorization/hypothesisation may, often be a chance outcome or the result of a self-revealing accident during the research process.

2.4.1 Hypothetico Deductive Approach

The Eleatics set out the hypothesis based on their opponent's arguments *and went ahead to prove it wrong* through (a) either deducing a conclusion which appeared absurd or impossible, or (b) deducing two conclusions incompatible with each other, *which again was not acceptable.* Socrates and Plato who followed them preferred to set out the hypothesis *for explaining* a phenomenon rather than disproving a point of view. This was motivated by *the rationalist approach* that saw a hidden law (and harmony) in nature based on the *relation between cause and effect.* This relationship was, moreover, thought to be unidimensional since *effect follows from cause,* which is same as *if 'X, then 'Y'*. And once, the underlying link between cause and effect was well understood, it enabled the knower the power to predict results/events.[3]

The hypothesis/premise set out in the beginning *may, nevertheless, employ several general laws/axioms on which there is agreement.* In this method of deduction, mathematics is a very helpful ally. Astronomy, for instance, calls upon mathematics to articulate reasons for the motions it observes. It has been, further, argued 'without the mathematical deduction which leads from the hypothesis *to a concrete testable situation*, there would be no knowing what to test' (Russell, 1959). In other words, conclusion so arrived at ought to be compared with facts.

2.4.2 Empirico –Inductive Approach

Under hypothetico-inductive approach the proposition/hypothesis is proved true or false, *empirically.* Quite often, the universe of the subject matter is very large and varied and it is not possible to study each and every state. "Surveys" in place of "census" may be the method preferred in the research investigation. The basic question in proving/disproving the hypothesis is the adequacy (validity) of evidence. In this context it is essential to differentiate between 'eliminative' and 'enumerative' induction.

'In eliminative induction, hypothesis are stated as "universal generalization", so that they can be falsified *by a single negative case*' (Sjoberg, 1968). In contrast, under enumerative induction, hypothesis can be proved true or false *through actual enumeration of the whole population.* 'The greater the specifity demanded the more difficult it becomes to formulate testable and valid universal hypothesis. Thus, we can more readily hypothesize that all societies have a family system than certain types of relationship among husband and wife and children hold for all social orders' (Sjoberg, 1968).

Furthermore, the 'methodology problems of induction, as a result, center on the ways of analyzing and explaining the notion of probability, which involves considerable conceptional difficulties and effects the theoretical status of induction…Initially the frequency theory of probability was preferred by the inductivists who treated the probability statements as empirical statements describing the number of times some event has happened relative to the number of times it might have happened' (Bose, 1995).

'However, when unbound series of events are considered, the problem of induction reappears as the problem of trying to estimate the long run or limiting frequency from a finite run and to do this the uniformity principle has to be invoked. In recognition of this problem, alternative conception of probability was proposed in which it is posed as a logical relation between evidence and conclusion. This view of probability was later developed into *confirmation theory*, where the aim was to provide logic of induction and rules for calculating the degree of confirmation'(Bose, 1995).

'Despite its apparent sophistication, this theory faces two related difficulties: first, because *a proposition might be confirmed by one set of evidence statement but disconfirmed by other sets*, the confirmationist must insist upon the requirement of *total evidence* before calculating the degree of confirmation of the proposition. Second, this system of inductive logic assigns zero probability to the laws formulated as unrestricted generalizations, since no finite amount of evidence can confirm such generalization of potentially infinite scope. Another alternative emerging out of the difficulties encountered in probabilistic interpretation of induction suggests that it is more appropriate to offer a *pragmatic justification*, implying that if inductive principles cannot be validated they can be 'vindicated'. However, it remains for the vindicationists to solve the comparative problem of selecting competing hypothesis and to show how eventual convergence in the long run can bear upon the short-run judgetment' (Bose, 1995).

2.4.3. Retroductive or Abductive Approach

According to C. S. Pierce *(1839-1914),* both deductive and inductive forms of logic are incomplete in providing a hypothesis. In his view, there has to be always a two way process between a hypothesis *which may have been arrived tentatively* through deduction and the observation of 'facts'. A particular kind of probable hypothesis usually compels the researcher to continue looking and the data forces necessary modification of the initial hypothesis in reverse or retroductively. This is like evolving/formulating the hypothesis based on 'abduction' or 'retroduction'.

This was also the method adopted by Johannes Kepler (1571-1630) who revisited the Copernican theory of the circular orbits of planets around the sun with his own record of observations. He found the theory of circular orbits did not save the appearances. He also noted that when the planets are closer to the sun they move faster.

This only showed that any hypothesisation arrived at deductively based on preconceived notions and divorced from observation, did not do justice to facts. Pierce described Kepler's effort in this direction as 'the greatest piece of retroductive reasoning ever performed'.

2.5 Theory and Assumptions

Theories are the outcome of hypothesis, which have been sufficiently modified retroductively, tested and found correct. Theories are said to have three essential features, namely, *the broad logical structure*, the proposition about the specific issue having *the quality of generalization and the assumptions*. Elaborating the relation between theories and assumptions *in the field of economics*, Lionel Robbins [1898-1984] observed, "The *propositions of economic theory* like all scientific theory are obviously *deductions* from a series of postulates' (Robbins, 1955). The postulates may be about production, distribution, exchange or consumption in the economy. The postulates, in turn, are often based on assumptions, such as, *ceteris paribus* ('other things remaining the same'). This makes the final proposition or the conclusion reached *simply unrealistic*.

It was therefore observed, 'Thus the economists, firstly regard only one part of man's nature, and treat him simply as a money making animal; secondly, they disregard the influence of custom and only take account of competition.... Such hypothetical laws, though leading only to rough conclusions *are yet useful* in giving us a point of view from which to observe and indicate the existence of strong over-mastering tendencies' (Toyanbee, 1884). According to Arnold Toyanbee also, *deduction* should be combined with the induction *method (historical approach)* to reach conclusions in subjects like economics. As he remarks, 'Adam Smith and Malthus, indeed, had historical minds; but the form of modern text-books is due to Ricardo, whose mind was entirely unhistorical'.

2.5.1 Model and Equations

An *economic model, for instance,* portrays the inter-relationship between different activities in the economy through simplifying the actual conditions by means of basic assumptions. It also has been argued that *a model need not essentially be mathematical*. 'If the model is mathematical, however, *it will usually consist of a set of equations designed to describe the structure of the model.* By relating a number of variables to one another in certain ways, these equations give mathematical form to the set of analytical assumptions adopted' (Chiang, 1974). Moreover, 'each of these equations contains a certain number of the unknowns and a certain number of the given variables and constants ' (Tinbergen, 1967).

The equations that form part of the model are generally of three types, namely, *definitional, behavioral or equilibrium equation.* A definitional equation is same as 'identities' whereby the alternate expressions have the same meaning (i.e., $P = R - C$, that is, profit is same as total revenue minus total cost). A behaviorial equation, on the other hand, specifies the manner in which the independent variable behaves in response to changes in the dependent variable (i.e., $C = 75 + 10\,Q$, that is, total cost is related to the quantity produced in a certain way). The third type of equations is relevant to models involving the notion of equilibrium ($S = I$, that is, intended saving is equal to intended investment at 'equilibrium output').

The models may similarly be a macro or a micro model, a dynamic or a static model etc. It has been further argued that 'the advantage of working with models represented by sets of equations lies in the decomposition of the process of research into logically arranged elements, *facilitating its being checked.* The somewhat loose talk about direct and indirect "effects" of a certain measure of economic policy, usual in verbal analysis, may easily forget on several of these effects'(Tinbergen, 1967).

2.6 Classification of Research

Broadly classified, research may be grouped under two heads: (a) basic research and (b) applied research. *Basic research* aims at a better understanding of a particular subject and the findings of research may not be of direct or immediate use. Basic research is, nevertheless, of great significance as it helps remove doubts and opens new vista of knowledge. *Applied research* on the other hand, is directed towards problem solving and is concerned with the desired interventions. It, thus, emphasizes upon developing actionable factors as against *simply testing theories*. It is also said to stress on predicting effects rather than finding causes.

Any research involves as a first step, review of the existing literature/knowledge on the subject matter. Next comes the way one organizes the study and the methodology of research. It is worth mentioning that different people may approach the same phenomenon differently. Thus, '*a medical doctor* might examine the effect of the tribe's fishing habits upon the people's diet, whereas *a geographer* could study the fishing patterns from the perspective of the group's adaptation to the physical environment' (Sjoberg, 1968).

2.6.1 Methodology of Research

Methodology usually refers to the procedures or technique of data collection. As in the example above, collection of data cannot be at random and has to be with reference to the purpose in-hand *or relevant to an active categorization.* The data may then be collected based on secondary sources or primary sources. Primary sources of data collection may either be in the form of census, surveys, participative inquiry or experiments. Virtual reality based on computer graphics are powerful display devices and is emerging as yet another very powerful tool in research methodology. It creates life like artificial world that can be perceived, explored and manipulated at will.

The methodology of research in turn, may, however, get influenced by the philosophical position one adopts. The 'positivists' proceed in their research investigation with the belief that the researcher finds out the reality objectively and is unbiased. The 'phenomenologist' or 'social constructionists', on the other hand, are apprehensive of the subjectiveness involved since *'facts are essentially facts-as-interpreted, (Bose, 1995).* A phenomenologist may, therefore, not approve of looking at the (social) reality through categories determined by the researcher and prefer recourse to ideas people themselves generate. Accordingly, 'the *"taken for granted"* aspects of people's lives can be brought into question and "hidden" issues can be brought into the public sphere in a relatively safe way' (Laws, 2003).

Summing Up : Submitting a Research Proposal.

Preparation of research proposal for 'funding' calls for writing a synopsis. The 'synopsis' states the objective and the methodology of research. The objective of research must be stated with respect to (a) how is the study going to add to the existing stock of knowledge and

(b) how will the research be of use to the society. Moreover in regard to (a), the proposal may provide a brief survey of literature on the subject supported by a bibliography. It should also bring into focus the questions that remain 'unanswered', which may be exploratory, explanatory or descriptive. The methodology of research, on the other hand, requires stating the procedures or technique of approaching the problem. This may include elucidating especial requirements such as, those of field survey or the facility of a laboratory. If these are not available locally, that should also be stated clearly.

Notes

1 The Indian case, however, presents a special case since epistemology constituted part of the religious system. 'Logic was forced into the scheme as a means of salvation by the contention that clear thinking and logical arguments were essential means to the highest bliss, and thus a religious basis was given to what was essentially a system of reasoning' (A.L.Basham, "Logic & Epistemology", in '*The Wonder That Was India'*, Rupa & Co., New Delhi, 2003).

In the absence of universal education and the printing technology as of now, religion especially in the Indian society kept alive the spirit of enquiry for those who were inquisitive. The majority, perhaps, went its own way meeting the necessities, merry making and the other vicissitudes of life.

2 This is very close to the philosophy of Buddhism which also believed/believes in the universe to be a flux of momentary but interdependent events (see Basham, A.L. op.cit.)

3 David Hume (1711-1766), however, questioned the rationalist approach and argued that what is thought to be the cause and effect relationship *may be the outcome of a particular habit of thinking.*

Annexure 2.1

Stochastic Enquiry and Testing of (statistical) Hypothesis

'When we derive a hypothesis from a sample and hope it to be true for the population as a whole, such a hypothesis is known as statistical hypothesis' (Prasad, 1973). The testing procedure (t-test) employed in such (non-experimental) empirical studies, begins by setting up two hypotheses that are mutually exclusive. The "null hypothesis" *implying no difference between the sample and the universe* is designated as H0; the "alternative hypothesis" implying *dissimilarity between the sample and the working universe* is designated as H1.

Subsequent steps involve finding sufficient evidence to reject the "null hypothesis" in which case the "alternative hypothesis" is accepted. The stress on *rejecting* the "null hypothesis" emanates from *the notion of falsifiability* according to which no theory can be proved 'true' with finality; although it can be proved 'false' with finality (Pepper, 1961).

There are, moreover, four outcomes possible while testing the hypothesis in a system of inductive statistics. Two of these are good and two are bad. If we reject H0, when in fact it is true, we have what is termed as type I error. The other possible error results when we fail to reject a 'null hypothesis' (H0) that is in fact incorrect. This is a type II error. Most of the time, however, greater attention is paid to type I error.

Equally important to recognize is the difference between drawing inferences for simply reaching a conclusion and *for taking a decision.* The decision maker has a stake in accepting or rejecting a hypothesis *on account of its direct impact on outcome*. The *pragmatic*, therefore, attempts to reduce the probability in occurrence of the error. This is done through appropriately fixing the *confidence level* for rejecting the 'null hypothesis'. A measure that is closely related to the confidence level is the *significance level* denoted generally as (alpha). It is same as 1 minus the confidence level. Thus, 95 per cent confidence level is same as 5 per cent significance level.

Chapter 3

Policy for Research Promotion

The Government of India accorded due recognition to *scientific research* soon after independence. It created the Department of Scientific Research under the direct charge of the Prime Minister as early as 1948. *This included research relating to development of atomic energy.* 'The main functions of the Department of Scientific Research were to coordinate research, give scientific advice to Government departments, finance adhoc scientific research in universities and research institutes, deal with adhoc research fellowships in subjects in applied science and carry out all work in connection with the International Council of Scientific Unions and the International Scientific Unions' (Bhatnagar, 1949). The Government of India, subsequently, adopted *the Scientific Policy Resolution in 1958* (Annexure3.1) and *the Technology Policy Statement* was announced later in 1983 (Annexure 3.2). The Science and Technology Policy, 2003 is the latest policy statement on scientific research by the Government (Annexure 3.3).

In the field of humanities and social sciences as well, *significance of research was duly recognized.* The Planning Commission, thus, set up the *Socio-economic Research Wing* at the outset of planning in India. Various centres of advanced learning and research such as the Delhi School of Economics (1949), the Indian Institute of Public Administration (1954), the National Council of Applied Economic Research (1956), the Institute of Economic Growth (1958) were soon established. A good many of these institutions had the Prime Minister as the President of the Institution. The Indian Statistical Institute (Kolkata) founded by Professor P.C. Mahalonobis in 1931 was accorded the status of "an institution of national importance" in 1959 by an Act of Parliament. Apart from research in statistics, applied mathematics and physics, the institution has been involved in research in the fields of population studies, economics and application of statistics in social sciences.

3.1. Policy Initiatives for R & D

The Government gave adequate thrust on promotion of R&D from the very beginning through various measures both implicitly and explicitly to make the country self-reliant. Its origin could be traced back to the Statement of Industrial Policy 1945, which recognized scientific and industrial research as one of the prerequisites of industrial progress and emphasized on maintaining a high standard. Subsequently, allocation of financial resources for the development of science & technology became an integral part of the Five Year Plan. The Government moved in 1958 the 'Scientific Policy Resolution' in the Parliament, which laid the broad framework of the aims and strategy for promoting scientific research in all its aspects, namely, pure, applied and educational. The initial emphasis was on creating the necessary technological infrastructure and on creating the scientific manpower for industrial research. Enactment of the Indian Patent Act in 1970 by the Parliament provided significant encouragement to domestic technology development activity. Drugs & pharmaceuticals industry and the food processing industry in particular benefited most from this Act, especially, so since the Indian Patent Act (1970) recognized 'process patenting' and did not grant product patents for these products. It also reduced the life of process patents from 14 to 7 years. It thus helped the domestic firms to reengineer the known products through alternative processes in a cost-effective way.

Major policy initiatives for development R&D in India over the years since 1945 may be summarized as shown in Table 3.1.

The Government established the National Committee on Science and Technology (NCST) in 1971 to formulate a comprehensive plan for the development of science and technology. Based on its recommendation, a scheme for the recognition of in-house R&D units was introduced for providing tax concessions and other benefits to in-house R&D and for successfully adapting imported technology, introducing new products (and processes) and for developing import substitutes. It provided fiscal and non-fiscal incentives such as exemption from Quantitative Restrictions (QR) for import of equipment, prototypes, raw material, components and consumables; concession on import tariffs; 100% tax deduction on expenditure incurred on approved scientific research programs; and excise duty exemption on products based on in-house R&D as well as preference

Table 3.1 : Major Policy initiatives in R & D

Year	*Policy Initiative*
1945	Statement of Industrial Policy, 1945
1958	Scientific Policy Resolution, 1958
1970	Indian Patents Act, 1970
1983	Technology Policy Statement, 1983
1985	R&D Cess Act, 1985
1988	Venture Capital Guidelines, 1988
1995	Technology Development Board Act, 1995
1999	Amendment in Indian Patents Act 1970
2000	New Millennium Indian Technology Leadership, 2000
2003	Science and Technology Policy, 2003

in import of technology and capital equipment. Accelerated depreciation on equipment and special investment allowance were also added by 1977 in the list of permissible incentives. Furthermore, royalties earned by the firms on account of export of indigenous technology was given complete exemption from tax and the royalty earned within the country was allowed 40% rebate.

The Government reiterated its commitment to promote domestic technological capability through the announcement of the Technology Policy in Statement (TPS), 1983. It accordingly stated, "*there shall be a commitment to ensure an adequate scale of investment in R&D for the absorption, adaptation and, wherever possible, improvement on and generation of new technology* making fullest use of overall national capabilities. For all practical purposes, the TPS 1983 remained the major policy document for promoting scientific research within the country during the pre-liberalization period.

The post liberalization period since 1991 have been witness to a number of new policy announcements in quick succession, such as the Technology Development Board Act (1995); amendment of Indian Patents Act 1970 (1999); 'Vision (2020)', New Millennium. Indian Technology Leadership (2000) and Science and Technology Policy (2003). Amendment in the Indian Patent Act (1970) in 1999 was unique in the sense it recognized the system of product patent in place of process patent enjoyed by the three sectors, namely food, pharmaceutical and chemical and increased the duration for all patents granted under the Act to 14 years. In a way the amendments had to be made for complying with the WTO (World Trade Organization) obligations. The potential effect of these changes on the innovative behavior of the Indian pharmaceutical and chemical industry has become the subject of a very lively debate.

3.1.1 Research Schemes for Innovation

The prime objective of research and development (R&D) during the earlier years was to ensure efficient adaptation and absorption of imported technology and to promote development of indigenous technology. This approach has undergone a significant change. The approach paper to the Tenth FYP (2002-2007) states *"The approach to technology will go beyond technology import, absorption, adaptation or assimilation . . . Technology will be used as a tool to give India a competitive position in the new global economy . . . increasing India's share in high-tech products, deriving value from technology led exports and export of technology will be given a major thrust."* The overall approach seems to be creating enabling environment for R&D in terms of improving technology infrastructure, enhancing industry-institution linkages, development of science & technology manpower and providing fiscal incentives for encouraging in-house R&D. In pursuance, the Science and Technology Policy (2003) also recognize the inevitability of the processes of globalization and liberalization of the Indian economy. Among other things, it argues for the need for reconstruction of academic scientific system, technology development through appropriate reward mechanism, measures to increase active involvement of industry in both basic and applied research and generation and management of intellectual property.

The various schemes floated by the Government of India in recent years are summarized in Table 3.2.

A marked difference in approach of the government can be seen between the pre and post liberalization periods. The pre-liberalization period revolved generally around adaptive R&D and support of the Government for the same. This foreign collaboration (FC) objective was achieved mainly through the (three) instruments of : (a) imposing conditionality on FC agreements, (b) development of technology

Table 3.2 : Research Schemes for Technology Department

Name of Scheme	*Scope*
1. Program aimed at Technological self-reliance (PASTER)Commenced: 1993 Implementation: DSIR	• Projects involving R&D and Engineering activities for: (a) absorption and up gradation of imported technology; (b) development and demonstration of new or improved technologies; • These projects concentrate on know-why, product and process development and improvement • The scheme also encourages involvement of GRIs (Government Research Institute) with Industry • The partial financial support by DSIR covers prototype development and pilot plant, testing and evaluation
2. Fund for Technology Development and Application (FTDA) Commenced: 1996 Implementation: TDB and DST	• Equity capital (up to 25 % of the project cost, including margin money for working capital, provided such investment does not exceed the capital paid-up by the promoters), or • Soft loans (at 6 % simple rate of interest) to enterprises, co-operatives and other agencies as applicable, attempting development and commercial application of indigenous technology, or adapting imported technology for widen domestic application;
3. Home Grown Technology Program (HGT) Commenced: 1993 Implementation: TIFAC	• It supports projects by Indian Industry to commercialize new technologies and to further excel in existing technology fronts by providing: (a) partial financial assistance upto 50 % of project cost and (b) techno-managerial assistance on soft terms. Repayment is spread over a period of 5 years after successful completion of the project.
4. Technopreneurs Promotion Program (TEPP) Commenced: 1998-99Implementation: DSIR and DST	• Activities under TEPP includes providing financial support to selected individual innovators having original ideas for converting them into working models, prototypes etc. • 90% of the total project cost is covered by the scheme

Table 3.2 (Contd....)

Name of Scheme	*Scope*
5. Technology Projects on Mission Mode Implementation: TIFAC	• Three schemes, namely: (a) Advanced composites, (b) Sugar production technology and (c) Fly Ash disposal and utilisation. • The project should generally be a tripartite collaborative effort involving industry partner, research laboratory/institution and TIFAC. • Assistance up to 50% of the total project cost in the form of Technology Development Assistance (TDA).

Sources : DSIR (2004); DST (2004); TDB (2004); TIFAC (2004); Mani (2002)

Note: PASTER is the only research grant scheme, while the other two are basically loan schemes though in the case of FTDA there is equity participation as well. There is certain amount of duplication between the various schemes e.g. the PASTER and FTDA are targeted at the same set of projects and are administered by sister departments.

infrastructure and (c) fiscal & non-fiscal incentives. The conditionality relating to progressive import substitution of components, restrictions on duration of agreements; encouraging firms to create component suppliers by way of retaining the right to sell or sublicense the imported technology; easier clearance to the firms that can demonstrate their in-house R&D capability as well as emphasis on training of personnel all acted synergistically to induce firms to undertake in-house R&D. In the present liberalized scenario, on the other hand, there is no compulsion of undertaking R&D for import substitution as the non-financial conditionality on FC as well as the import policy for the same have been relaxed considerably. The driving force for in-house R&D today is the need for survival in a highly competitive market. The main instruments for inducing in-house R&D are, therefore, provision of adequate institutional support and the fiscal & non-fiscal incentives.

3.2. Fiscal Incentives for Scientific Research

The main fiscal incentives that exist today for promoting R&D in India are mentioned below:

Direct Tax Incentives:

- 100% write off of revenue and capital expenditure for purpose of tax liability out of the profits/income, for in-house R&D. Deduction for capital expenditure has been raised to 150% for certain specified industries such as drugs & pharmaceuticals, electronic equipment, telecom equipment industry etc.
- Weighted tax deduction of 125% for the financial contribution by the firms for sponsored research in approved national labs, universities etc.
- Income-tax exemption of 125% of donations made to approved scientific and industrial research organizations,
- Accelerated depreciation allowance for investment on plant & machinery based on indigenous technology,
- 10 years tax holiday to the firms created exclusively for R&D activities.

Indirect Tax Incentives:

- Exemption from Custom duty for recognized in-house R&D units on import of capital equipment, instruments, spares, accessories and consumables for R&D activities,
- Exemption from Custom duty on imports made for R&D projects funded by the government in industry,
- Exemption from Excise duty for 3 years on goods designed and developed by a wholly owned Indian company and patented in any of the two countries out of India: USA, Japan and European Union.

Fiscal incentives, both direct and indirect, are discussed in greater detail in the paragraphs below.

a. Tax Holiday : Section 80-IB(8A) of the Income Tax Act, 1961 provides for ten year tax holiday to companies engaged in (commercial) research and development. The company should, however, have the due approval for undertaking R&D activity from the prescribed authority namely the Secretary, Department of Science and Industrial Research (DSIR).
b. Customs Duty Exemption : All Scientific and Industrial Organization (SIRO), recognized by DSIR are eligible for customs duty exemption on input of scientific equipment, instruments,

spares accessories as well as consumables for research and development etc. Essentiality certificate are issued to the applying organizations by the Screening Committee set up by DSIR.

c. Excise Duty Exemption : All Scientific and Industrial Reearch Organizations (SIRO) recognized by DSIR are eligible for excise duty exemption on the purchase of scienticfic and technical instruments, apparatus, equipment (including computers) accessories and spare parts thereof and consumables; computer software micro-films etc. The Screening Committee set up by DSIR issues essentiality certificate to the applying organizations.

d. Depreciation Allowance on Plant & Machinery : Accelerated depreciation in respect of plant and machinery engaged in R&D is allowed. Secretary, DSIR, Ministry of Science and Technology is the prescribed authority to certify expenditure where higher rate of depreciation is to be allowed. All such applications received are examined in the Department based on visits to plants to verify the claims.

e. Writing off of Capital Expenditure : Section 35(3) of IT Act, 1961 provides inter-alia, that expenditure made on equipment and related to research activities may be written off 100% out of the profits/income for purpose of direct taxes in the year in which the expenditure are incurred. In cases of any dispute, the Director General (Income Tax Exception) is the prescribed authority to settle disputes which he does based on the communication/ recommendation received from the Secretary, DSIR.

f. In-House R&D : 'Finance Bill 1997 introduced sub-section 2 (AB) in Section 35 of the Income Tax Act, 1961 to encourage R&D in drugs, pharmaceuticals, electronic equipment, computers, telecommunication equipment and chemicals. The Finance Act, 2000, further enhanced the tax benefits. Accordingly, the weighted tax deduction was raised to 150% of the expenditure incurred. The bill was further amended in 2001 to include expenditure on in-house R&D for units engaged in biotechnology.

g. Tax Exemption to Public Funded Institution : 'Public funded research institutions, universities IITs, IISc (Bangalore) Regional Engineering Colleges are eligible for availing customs duty exemption on import of equipment, spares and accessories. They are also eligible for central excise duty waiver on purchase of indigenously manufactured items for scientific research purposes. An inter-departmental screening committee under the DSIR is the empowered body for recommending the registration, which is valid for a maximum period of five years.

3.3. Venture Capital and R & D

Since developing a product or a process is a risky proposition and may not be a success in the market, there are not enough people/ entrepreneurs around to invest in research and development (R&D). *Lack of funds* or 'capital gap' with the innovators who have the vision and 'the concept' is another handicap in stimulating innovations. Venture Capital Fund (VCF) has emerged as an answer to these unconventional, novel and risky propositions. Amongst, the successful enterprises, often quoted, which benefited from venture capital fund, are Kodak (colour photos), Microsoft, Intel and Apple Computers.

The Government of India enacted the R&D Cess Act in 1986 to generate funds for VCF. Accordingly, a cess is levied on all payments made for imports of technology at the rate of 5 per cent of the value of imported technology. The cess, which was originally operated by the IDBI is now being managed by the Technology Development Board. Some of the other financial institutions, foreign institutional investors (FIIs) and banks like the ICICI, SBI and Canara Bank have also set aside certain corpus of fund to be used as venture capital. The number of active members of the Indian Venture Capital Association (IVCA) had gone up to 33 by 1997.

The Securities and Exchange Board of India (SEBI) later came out with Venture Capital Fund Regulation in 1996. According to the Regulations, a Venture Capital Fund (VCF) may be structured either as a company registered under the company Act, 1956, or as a trust under the Indian Trust Act and registered under the Indian Registration Act, 1908. Subsequently, they could apply to SEBI for grant of permission for carrying out venture capital activity. SEBI may, however, ask from the applicants to furnish such information, as it considers necessary for the grant of the certificate (permission).

Under the Regulations, a venture capital fund may be raised from any source whether Indian, foreign or non resident Indian, which should ordinarily be not a value less than Rs.500,000/-. Furthermore, no unit financed by the VCF can be listed on the stock exchange before the expiry of three years (that is, unless it has proved itself/succeeded). Companies (banks) or trusts financing such (R&D) start-up ventures may provide the requisite finance in the form of *equity* or *loan*. In the case of equity, the attraction is that if the enterprise is successful, the worth of the equity jumps sky high in no time. The financier's stakes are, however, great in case of possible failure; hence, they sit on the Boards of such venture companies, take interest in product design and help in marketing plans.

Income Tax Benefits

The Income Tax Act (1961) exempts the income/profit for payment of tax for three assessment years in start-up ventures through Venture Capital Fund (VCF), from investments made in the form of equity or loan. This exemption, is however, allowed to specific sectors, such as, information technology, basic drugs in pharmaceuticals, bi-technology and agriculture and allied sectors or such sectors the Government of India decides to notify. *The SEBI Regulations, however, do not prescribe the industries or sectors in which the VCF has to operate*. The exemptions are also allowed for the production of any article or substance for which *patent has been generated* by the national research laboratory (or any other scientific research institution approved by the Department of Science and Technology, Government of India). In the context of R&D, this could perhaps be modified to also include *development of products and processes*.

Venture capital fund is emerging as an important source of finance for R&D in India. The venture capital fund grew from around $ 20 million in 1996-97 to $ 1200 million by 2000-01. Firms like Avesthagen, Strand Genomics, Bharat Biotech, Shantha Biotechnics, Lttiam, Tejas Networks, Ishoni Networks, Mitoken, etc. are the new venture assisted firms that have innovated new products or services. Some of the firms like Lmpulsesoft, Tejas Networks, Reva and Lttiam have become product developers for the global market. The Technology Development Board, on the other hand, has been effective in stimulating innovations in India in areas such as Statins (Biocon), Electric car (Reva Electric Car), Hepatitis B Vaccine (Shantha Biotechnics), Rota-Virus Vaccine (Bharat Biotech), Lithium ion batteries (21st Century Batteries) and High strength alloys (AV Alloys) (ASCI, 2003). It has been further observed, 'India has become one of the most preferred destinations of off-shore venture funds, after Japan' (ASCI, 2003).

3.4. Foreign Direct Investment and R & D

Foreign Direct Investment (FDI) in R&D are routed through the Foreign Investment Promotion Board (FIPB) under the Department of Industrial Policy and Promotion, Ministry of Commerce and Industry. The FIPB is a committee of Secretaries with representation from Ministry of Finance, Ministry of External Affairs, Ministry of Small Scale Industries and Department of Commerce. (Investment exceeding Rs.500 crores are, however, required to get the approval of the Cabinet Committee on Foreign Investment (CCFI). The FIPB considers applications on the basis of notified guidelines and disposes them within a 6-8 week time frame as has been laid down by the Cabinet. The entire process of FIPB applications, starting from registration to listing on FIPB agenda and their final disposal and dispatch on official communication is placed on the website for transparency in decision making and to enhance investor confidence.

Once the project is approved the Foreign Investment Implementation Authority (FIIA) helps them on obtaining the required clearance. The Secretary, Department of Industrial Policy and Promotion chairs the FIIA. The FIIA framework has been further strengthened with the adoption of the six point strategy which includes, inter-alia (a) close interaction with companies at both operational and board room level (b) follow up with administrative ministries/State Governments and other concerned agencies and (c) sector specific approach to help solve the problems.

3.4.1. Foreign Technology Agreements and R&D

According to the Steering Committee on Foreign Direct Investment (Planning Commission), India has one of the most liberal policy regimes in regard to technology agreements. Foreign technology collaboration are permitted both through the automatic (RBI) route and the FIPB route.

RBI accords *automatic approval* for all industries for foreign technology collaboration agreements subject to :

a. The lump sum payments to foreign collaborators not exceeding US$ 2 million

b. The royalty payable is limited to 5 per cent for domestic sales and 8 per cent for exports subject to total payment of 8 per cent on sales over a 10-year period.

c. The period for payment of royalty does not exceed 7 years from the date of commencement of commercial production or 10 years from the date of agreement whichever is earlier.

For the following categories, Government approval is necessary under the FIPB route:

i. Proposals attracting compulsory licensing.

ii. Items of manufacture reserved for the small-scale sector.

iii. Proposals involving any previous joint venture or technology transfer/trade mark agreement in the same or allied field in India.

iv. Extension of foreign technology collaboration agreements (including those cases, which may have received automatic approval in the first instance).

v. Proposals not meeting any or all of the parameters for automatic approval.

Proposals needing compulsory licensing are forwarded by the Secretary of Industrial Approval (SIA), Department of Industrial Policy and Promotion to the Screening Committee of the Department of Scientific and Industrial Research, Ministry of Science and Technology. Based on their recommendations, the Letter of Intent (LOI) is issued to either the Indian entrepreneurs/NRIs or foreign entrepreneurs for carrying on business.

Moreover, the different components of foreign technology collaboration such as technical know-how fees, payment for design and drawing, payment for engineering service and royalty *are eligible for approval both through the automatic route and by the FIPB Government route*. Payments for hiring of foreign technicians, deputation of Indian technicians aboard, and testing of indigenous raw material, products, indigenously developed technology in foreign countries are, however, governed by separate RBI procedures and rules and are not covered by the foreign technology collaboration approval. Similarly, payments for imports of plant and machinery and raw material are also not covered by the foreign technology collaboration, for which RBI is the competent authority to give the approval.

3.5 Institutions of Higher Learning and Research

The National Policy of Education (1986) enunciated on the eve of the Eighth Five Year Plan included the policy guidelines for research studies in India. The NPE observed, 'there is a symbiotic relationship between higher education and research. Research cannot be carried out without the vital support of higher education and education would be dull and monotonous without the opportunity and inputs of creativity'. It especially laid stress on the pre-eminent role of universities and institutions of higher learning in the field of basic research. At the same time it recognized that research, with all its economic/ intellectual returns, requires systematic flow of funds and management structures. *It also recommended for creating cooperative research facilities within the system of higher education for capital-intensive projects.* If need be, it argued, national research facilities should be set up within the system of higher education with the desired autonomy.

3.5.1. Science & Technology and Industry

One of the important recommendations of the NPE was the suggestion *for encouraging public and private sector enterprises to set up at least some of their R&D activities on the campuses of higher education.* As it observed 'each major industrial organization should be encouraged to develop strong cooperation and links with one or two institutions so as to utilize them intensively for research and development'. This policy recommendation appears to have been commendably abided by in the research and development programmes of the All India Council of Technical Education (AICTE) under the Ministry of Human Resources Development, such as, *the Industry-Institute Partnership Programme (IIPP) and the National Facilities in Engineering and Technology with Industrial Collaboration (NFETC).* The NPE also advocated for encouraging public sector research organizations such as the CSIR and the DRDO to implement some of their R&D activities through universities. It was felt this would provide a shot in the arm for university education and also provide cost-effective results for the sponsors.

3.5.2. Research in Social Sciences

Good quality research in social sciences is of equal importance. The NPE (1986), however, commented that in spite of a large number of universities imparting teaching in social sciences, *there are not many centres of real excellence in these fields*. It also observed, 'there is also a need for dovetailing of research findings into the syllabi and for linkages between social science research and other sectors of research. There is a real danger of creating a one-dimensional man in the absence of such linkages'. Management institutes such as the IIMs (Indian Institutes of Management), the NCAER (National Council of Economic Research), the NIRD (National Institute of Rural Development) and the ISIs (Indian Statistical Institutes) do enjoy good reputations and are generally full with requests for research studies. Amongst the private research organizations, the CMIE (Centre for Monitoring Economic Intelligence), the ICRA (Investment Information and Credit Rating Agency), CPR (Centre for Policy Research), the ICRIER (Indian Council for Research on International Economic Relations), the ORG (Operation Research Group), the TCS (Tata Consultancy Research) and similar organizations enjoy very good reputations in social sciences research. Amongst the multinational research organizations active in India, the McKinsey & Company and A. T. Kearney, Inc have made a mark in socio-economic research.

3.5.3 National Accreditation and Assessment Council

The UGC set up the National Accreditation and Assessment Council (NAAS) in 1994, as an autonomous body with its headquarters in Bangalore. Its main function is that of accreditation of universities and colleges on a five-point scale. Evaluation is voluntary and is based on 'self assessment appraisal'. The philosophy of NAAC is inspired by *objectivity* and *for improving standard of education* rather than judgmental and punitive. Self-assessment, in turn, is based on teaching learning, student support and progression, *research and consultancy* and infrastructure facilities. The certification so achieved is considered valid for five years. Accreditation benefits the institution in identifying internal areas of planning and resource allocation, gives institutions a new sense of identity, enhances collegiability on the campus since everyone matters in building the image, provides funding agencies with objective data for sponsoring studies and enables society/ employers with reliable information on the quality of education.

There are nevertheless lessons to be learnt from United Kingdom (UK). The privatization drive in the UK in the 1980's impacted the universities as well. As such if there were no restriction on financial assistance, there was nothing to worry for the colleges and the university departments. However, if resource constraint surfaced *then the faculty with least research output to show had to leave.* Research output in each discipline in U.K., is evaluated by the Select Panel of experts constituted (privately) by the various Academic Associations. Hence, the Royal Economic Society constitutes the Select Panel for evaluating the quality of research for the Economic Department in different universities/colleges. Evaluation is done on a five-point scale beginning with five stars and ending with one star. In view of the fame and reputation of the members of the Select Panel the certification is sufficient for the sponsors for deciding whom to request for the research project.

Teaching and research, moreover, go hand-in-hand with eight months of teaching and four months of research. In such a case, there is no relevance of schemes like Research Awards for teachers. It has nevertheless been pointed out that good research is, very often, the result of interface with other members of the faculty. As such, if the department strength were too small in the range of four to five members only, that would itself work as a limiting factor. One-way to overcome this thinly spread faculty *is to have fewer Colleges/Universities and enlarge the faculty* strength. Alternatively, multi-disciplinary research may be encouraged in colleges and universities as a policy, wherever the department strength is found to be small.

3.5.4 Shortcomings Highlighted by National Policy Education (NPE)

Some of the major problems highlighted by the NPE (1986) for R&D in higher educational institutions are mentioned below:

1. Most of the effort is concentrated in a few institutions and not spread over the entire system. This should be seen in the context of the need to grow R&D culture in all institutions and at the same time using selectivity for creating and support groups of excellence.
2. Experimental scientists and engineers have the greatest handicap in the educational sector due to poor infrastructural facilities. Maintenance, running cost and technical support is not adequately provided for. Instrumentation is generally poor in the educational institutions.

3. An atmosphere and pace appropriate to research is difficult to sustain in the face of cumbersome rules, procedural difficulties and most of all, lack of resources.
4. Cooperation and collaboration between research centres located in universities with R&D laboratories is scare.
5. There is little premium on quality and excellence especially in the area of relevant research and development. *An appropriate system for recognizing excellence and achievements and rewarding performance is yet to evolve.*
6. Mediocrity in R&D has to be guarded against. Research is by and large confined to traditional areas; inter disciplinarily has not taken root. *Choice of research problems is often dictated by considerations of easy publication in journals.*
7. Even in other wise well-equipped laboratories adequate provisions for spares, consumables and replacement of short-lived equipment is not available. Overheads need to be provided to the institutions by agencies sponsoring research. Without such overheads the institutions find it difficult to support research.
8. Poor library, inadequate information system, absence of computational and reprographic facilities are endemic to the majority of educational institutions. The need for modernization and removal of obsolescence in research facilities is urgent.
9. Very little research is undertaken to improve existing technology or applying existing or newer technologies to meeting the needs of villages and rural areas. Generally technologies, which increase productivity and improve the performance of an existing process, are not being developed.

It was also suggested that the UGC should have within its own system, *a central committee* looking at the R&D activities in the university system, just as *an advisory committee* for the technical education wing of the Ministry of HRD, should keep a close watch on *the modernization of laboratory facilities, facilities of good libraries and the like*. The NPE (1986) accordingly observed, 'such Committees should ensure that the kind of R&D projects taken up in the education sector are not only of importance to the concerned fields but also have some bearing on the national needs'. On the front of libraries, however, it appears very little progress could be made so far.

Annexure 3.1

Scientific Policy Resolution, 1958

The key to national prosperity, apart from the spirit of the people, lies in the modern age, in the effective combination of three factors, technology, raw materials and capital, of which the first is perhaps the most important since the creation and adoption of new scientific techniques can, in fact, make up for a deficiency in natural resources, and reduce the demands on capital. But technology can only grow out of the study of science and its applications.

The dominating feature of the contemporary world is the intense cultivation of science on a large scale, and its application to meet a country's requirements. It is this, which, for the first time in man's history has given to the common man in countries advanced in science, a standard of living and social cultural amenities, which were once confined to a very small privileged minority of the population. Science has led to the growth and diffusion of culture to an extent, never possible before. It has not only radically altered man's mental horizon. It has thus influenced even the basic values of life, and given to civilization a new vitality and new dynamism.

It is only through the scientific approach and method and the use of scientific knowledge that reasonable material and culture amenities and services can be provided for every member of the community, and it is out of a recognition of this possibility that the idea of a welfare state differs widely from the country to country indirect relation to the extent of industrialization and the effort and resources applied in the pursuit of science.

The wealth and prosperity of a nation depend on the effective utilization of its human and material resources through industrialization. The use of human material for industrialization demands its education in science and training in technical skills. Industry opens up possibilities of greater fulfillment for the individual. India's enormous resources of manpower can only become an asset in the modern world when trained and educated.

Science and technology (S&T) can make up for deficiencies raw materials by providing substitutes or indeed by providing skills which can be exported in return for raw materials. In industrializing a country, a heavy price has to be paid in importing science and technology in the form of plant and machinery, highly paid personnel and technical consultants. An early and large-scale development of science and technology in the country could therefore greatly reduce the drain on capital during the early and critical stages of industrialization.

Science has developed at an every-increasing pace since the beginning of the country, so that the gap between the advanced and backward countries has widened more and more. It is only by adopting the most vigorous measures and by putting forward our utmost effort into the development of science that we can bridge the gap. It is an inherent obligation of a great country like India, with its tradition of scholarship and original thinking and its great culture heritage, to participate fully in the march of science, which is probably man-kind's greatest enterprise today.

The Government of India have accordingly decided that the aims of their scientific policy will be:

1) to foster, promote and sustain, by all appropriate means, the cultivation of science, and scientific research in all its aspects – pure, applied, and educational;
2) to ensure an adequate supply, within the country, of research scientists of the highest quality, and to recognize their work as an important component of the strength of the nation;
3) to encourage, and initiate, with all possible speed, programmes for the training of scientific and technical personnel, on a scale adequate to fulfill the country's needs in science and education, agriculture and industry and defence;
4) to ensure that the creative talent of men and women is encouraged and finds full scope in scientific activity;
5) to encourage individual initiative for the acquisition and dissemination of knowledge, and for the discovery of new knowledge, in an atmosphere of academic freedom;
6) and, in general to secure for the people of the country all the benefits that can accrue from the acquisition and application of scientific knowledge.

The Government of India have decided to pursue and accomplish these aims by offering good conditions of service to scientists and according them an honoured position, by associating scientists with the formulation of policies, and by taking such other measures as may be deemed necessary from time to time.

Annexure 3.2

Technology Policy Statement, 1983

Political freedom must lead to economic independence and the alleviation of the burden of poverty. We have regarding science and technology as the basis of economic progress. As a result of three decades of planning, and the Scientific Policy Resolution of 1958, we now have a strong agricultural and industrial base and a scientific manpower impressive in quality, numbers and range of skills. Given clear-cut objectives and the necessary support, our science has shown its capacity to solve problems.

The frontiers of knowledge are being extended at incredible speed, opening up wholly new areas and introducing new concepts. Technological advances are influencing life-styles as well as social expectations.

The use and development of technology must relate to the people's aspirations. Our own immediate needs in India are the attainment of technological self-reliance, a swift and tangible improvements in the conditions of the weakest sections of the population and the speedy development of backward regions. India is known for its diversity. Technology must suit local needs and to make an impact on the lives of ordinary citizens, must give constant thought to even small improvements, which could make better and more cost-effective use of existing materials and methods of work. Our development must be based on our culture and personality. Our future depends on our ability to resist the imposition of technology which is obsolete or unrelated to our specific requirements and of policies which tie us to systems which serve the purposes of others rather than our own, and on our success in dealing with vested interests in our organizations; governmental, economic, social and even intellectual, which bind us to outmoded systems of institutions.

Technology must be viewed in the broadest sense, covering the agricultural and services sectors along with the obvious manufacturing sector. The latter stretches over a wide spectrum ranging form village, small-scale and cottage industries (often based on traditional skills) to medium, heavy and sophisticated industries. Our philosophy of a mixed economy involves the operation of the private, public and joint sectors, including those with foreign equity participation.

Our directives must clearly define systems for the choice of technology, taking into account economic, social and cultural factors along with technical considerations; indigenous development and support to technology, and utilizations of such technology; acquisition of technology through import and is subsequent absorption adaptation and up gradation; ensuring competitiveness at international levels in all necessary areas; and establishing links between the various elements concerned with generation of technology, its transformation into economically utilizable form, the sector responsible for production(which is the user of such technology), financial institutions concerned with the resources needed for these activities, and the promotional and regulating arms of the Government.

The Technology Policy Statement is in response to the need for guidelines to cover the wide-ranging and complex set of inter-related areas. Keeping in mind the capital scare character of a development economy it aims at ensuring that our available natural endowments, especially human resources, are optimally utilized for a continuing increase in the well being of all sections of our people.

We seek technological advancement not for prestige or aggrandizement but to solve our multifarious problems and to be able to safeguard our independence and our unity. Our modernization, far from diminishing the enormous diversity of our regional traditions should help to enrich them and to make the ancient wisdom of our nation more meaningful to our people.

Our task is gigantic and calls for close coordination between the different departments of the Central and State Governments and also of those concerned, at all levels, with any sector of economic, scientific or technological activity, and, not least, the understanding and involvement of the entire Indian people. We look particularly to young people to bring a scientific attitude of mind to bear on all our problems.

Aims and Objectives

The basic objectives of the Technology Policy will be the development of indigenous technology and efficient absorption and adaptation of imported technology appropriate to national priorities and resources. Its aims are to:

a) attain technological competence and self-reliance to reduce vulnerability, particularly in strategic and critical areas, making the maximum use indigenous resources;

b) provide the maximum gainful and satisfying employment to all strata of society with emphasis on the employment of women and weaker sections of society;

c) use traditional skills and capabilities, making them commercially competitive;

d) ensure the correct mix between mass production technologies and production by the masses;

e) ensure maximum development with minimum capital outlay;

f) identify obsolescence of technology in use and arrange for modernization of both equipment and technology;

g) develop technologies which are internationally competitive, particularly those with export potential;

h) improve production speedily through greater efficiency and fuller utilization of existing capabilities and enhance the qualify and reliability of performance and output;

i) reduce demands on energy, particularly energy from non-renewable sources;

j) ensure harmony with the environment, preserve the ecological balance and improve the quality of the habitat; and

k) recycle waste material and make full utilization of by-products.

Self-Reliance

In country of India's size and endowments, self-reliance is inescapable and must be at the very heart of technological development. We must aim at major technological breakthroughs in the shortest possible time for the development of indigenous technology appropriate to national priorities and resources. For this, the role of different agencies will be identified, responsibilities assigned and the necessary linkages established.

Strengthening the Technology Base

Research and Development, together with science and technology education and training of a higher order, will be accorded pride of place. The base of science and technology consists of trained and skilled manpower at various levels, covering a wide range of disciplines, and an appropriate institutional, legal and fiscal infrastructure. Consolidation of the existing scientific base and selective strengthening of the technology base in newly emerging and frontier areas such as information and mat4erials science, electronics and bio-technology. Education and training to upgrade skills are also of utmost importance. Basic research and the building of centers of excellence will be encouraged.

Skills and skilled workers will be accorded special recognition. The quality and efficiency of the technology generation and delivery systems will be continuously monitored and upgraded. All of this calls for substantial financial investments and also strengthening of the linkages between various sectors (educational institutions), R&D establishments, industry and governmental machinery.

Priorities: Need for Perspective Planning

The time scales involved in the generation of technology are long, even with imported elements. Therefore, relevant technologies in all areas of priority, particularly where large investments are to be made, should be clearly identified well in advance. The cost and time element involved in the import of technology and indigenous development will be given consideration. Components, which could be assigned to the various institutions which are capable of developing them or which could be built up for such activities, will be identified. Ministries concerned with large investments and production activities in as such as food, health and energy will be provided with appropriate technical support through suitably structured S&T groups.

Employment

Human resources constitute our richest endowment. Conditions will be created for the fullest expression and utilization of scientific talent. Measures will be taken for the identification and diffusion of technologies that can progressively reduce the incidence of poverty and unemployment, and of regional inequalities. The application of science and technology for the improvement of standards of living of those engaged in traditional activities will be promoted, particularly household technologies. Technologies relevant to the cottage, village and small industries sector will be upgraded. In the decentralized sector labour must be diversified and all steps taken to reduce drudgery. In all sectors, the potential impact on employment will be an important criterion in the choice of technology.

Energy

Energy constitutes an expensive and sometimes scarce input. Therefore, the energy requirements both of a direct and indirect nature for each product and each production activity and the associated technology employed will be analysed. Measures will be devised to avoid wastage or non-optimal use of energy. Fiscal measures as necessary will be introduced to ensure these. Research and Development in the energy sector will aim at improving the efficiency of its production, distribution and utilization as well as improvement of efficiency in process and equipment.

Technologies already employed will be evaluated on a continuing basis to raise maximum benefits in terms of increased production and lower costs, especially in the public sector enterprises. Every effort should be made to utilize by-products and wherever possible to recycle waste materials, especially those from urban areas. Programmes to make use of easily available and less costly materials will be supported.

Environment

Development should not upset the ecological balance for short as well as long-term considerations. Poorly planned efforts to achieve apparently rapid development, ignoring the long-term effect of many technologies on the environment, have resulted in serious ecological damage. It is, therefore, essential to analyse the environmental impact of the application of each technology. Due regard will be given to the preservation and enhancement of the environment in the choice of technologies. Measures to improve environmental hygiene will be evolved.

Some specific Areas

In technology development special emphasis will be focused on food, health, housing, energy and industry. In particular stress will be laid on:

- Agriculture including dry land farming;
- optimum use of water resources, increased production of pulses and oilseeds;
- provision of drinking water in rural areas, improvement of nutrition, rapid reduction in the incidence of blindness, eradication

of the major communicable diseases (such as leprosy and tuberculosis), and population stabilization;

- low-cost housing; development and use of renewable non-conventional sources of energy; and
- industrial development.

Indigenous Technology: Importance of Technology Development

Fullest support will be given to the development of indigenous technology to achieve technological self-reliance and reduce the dependence on foreign inputs, particularly in critical and vulnerable areas and in high value added items in which the domestic base is strong. Strengthening and diversifying the domestic technology base are necessary to reduce imports and to expand exports for which international competitiveness must be ensured.

The spirit of innovation and invention is the driving force behind all technological change. We must awaken our science and technology to the exciting challenges of our times, provide incentives to encourage inventors and direct their efforts to areas of special importance, the system of rewards and incentives will be strengthened for inventions, innovations and technological breakthroughs and their utilization. The fullest opportunity will be provided to make use of inventions.

Enhancing Traditional Skills and Capabilities

Traditional skills and capabilities will need to be upgraded and enhanced, using knowledge and techniques generated by advances in science and technology. Technologies, which will result in low-cost production and in products marketable close to the point of manufacture, particularly in the rural sector, will be promoted. Support will be given to technologies which reduce pressure on items in short supply and utilize improve local materials and methods. Government will give preference to products of such technologies in its own purchases. The adoption of technologies that can promote decentralized production will be helped through the support to design, marketing, qualify control and other services.

Ensuring Timely Availability

The time cycle from scientific research to utilization is a long one. Hence the need to initiate action well in advance to identify and ensure timely availability and delivery of new technologies. Encouragement and support (fiscal, commercial and administrative) will be given to the production and user organizations to be associated with and participate in technology development efforts at appropriate stages.

Upgradation to Prevent Obsolescence

Technology is constantly on the move. The base of indigenous technology should be capable of utilizing worldwide advances and adapting them to local needs. The creation and strengthening of institutional structures for keeping track of international development will receive urgent attention.

A strong central group will be constituted to undertake technology forecast and technology assessment studies, and will inter alia draw up programmes of purposeful research. Arrangements will be made to provide high-level scientific advice in major sectors of the economy. Where big investments are involved or a large number of volumes of production is envisaged, it will be incumbent on the Ministry or agency concerned to provide a technology forecast covering its requirements over a ten-year or longer period and evolve a strategy for development based on priorities.

Increasing the Demand for Indigenous Technology

Our country has already invested significant amounts in setting up research and development facilities as well as design consultancy and engineering capabilities. The technological potential inherent in this system of interlinked capabilities must be fully utilized and in turn provide a fillip for further development from within the system. Incentives will, therefore, be provided to users of indigenously developed technology and for products and processes resulting for such use.

Preferential Treatment

In view of the cost of technology development and the time necessary for successful marketing of a new or improved product, indigenously developed items are invariably at a disadvantage compared with imported products or those based on imported technologies and brand names. Support must therefore be provided through fiscal and other measures, for a limited period, in favour of products made through indigenously developed technologies, care being taken to ensure quality.

Fiscal Incentives

Suitable financial mechanisms will be established to facilitate investment on pilot plants, process demonstration units and prototype development in order to enable rapid commercial exploitation of technologies developed in laboratories. Linkages between scientific and technological institutions and development banks will be strengthened. Gaps in technology will be identified and suitable corrective measures taken with adequate allocation of resources. Fiscal incentives will be provided in particular to promote inventions; increase the use of indigenously developed technology; enhance in-house Research and Development in industry; and efforts directed to absorb and adapt imported technology.

Design Engineering

Capabilities in design engineering are essential for the translation of know-how to commercial production. This is particularly important in areas relating to: agricultural production; agro-industries; metallurgical, chemical and petrochemical processes'\; machine tools; industrial machinery and capital goods; as well as for the construction and erection of entire plants. Building up and enhancing these capabilities will have a catalytic beneficial impact on the utilization of indigenous efforts that have resulted in product and process know-how. Existing design engineering capabilities will be strengthened and upgraded and interaction encouraged between design engineering organizations, academic and research institutions and industry. Wherever gaps exist, design engineering capabilities will be developed and nurtured.

Engineering Consultancy

Engineering consultancy is a vital area for ensuring speedy technological and industrial development. It ensures the appropriate utilization of indigenous materials, plant and machinery. Engineering consultancy provides an essential link between R&D institutions and industry, and thus promotes effective transfer of technology. Capability for total systems engineering, process development and project management should be developed with collaboration if required. Wherever capability exists, utilization of Indian consultancy engineering organizations will be promoted. Even where foreign technical collaboration or consultancy is considered unavoidable, association of designated Indian consulting engineering organizations would be preferred. Indigenous engineering consultancy, in both private and public sectors, will be promoted on a found professional basis in the context of the overall national perspectives of technological self-reliance.

In-house R&D

In house R&D units in industry provide a desirable and essential interface between efforts within the national laboratories and the educational sector as well as production in industry. Appropriate incentives will be given to the setting up of R&D units in industry and for industry including those on a cooperative basis. Enterprises will be encouraged to set up R&D units of a size to permit the accomplishment of major technological tasks.

Technology Acquisition (Mix of Indigenous and Imported Technology)

A policy directed towards technological self-reliance does not imply technological self-sufficiency. The criterion must be national interest. Government policy will be directed towards reducing arrangements dependence in key areas.

Advantage should be taken of technological developments elsewhere. This can also be achieved through well-defined collaborative arrangements in research and development.

At any given point of time, there will be a mix of indigenous and imported technology. However, technology acquisition from outside shall not be at the expenses of national interest. Indigenous initiative must receive due recognition and support.

In the acquisition of technology, consideration will be given to the choice and sources of technology, alternative means of acquiring it, its role in meeting a major felt need, selection and relevance of the products, costs and related conditions. A National Register on Foreign

Collaboration will be developed to provide analytical inputs at various stages of technological acquisition.

Principles of Acquisition and Technology Assessment

Where the need to import technology is established, every effort should be made to ensure that it is of the highest level, consistent with requirements and resources. The technology import will be so planned as to have effective transfer of basic knowledge (know-why) and to facilitate further advancement.

Where the import of technology is contemplated, the level to which technology has been developed, or is in current use, within the country, shall be first evaluated. Lists of technologies that have been adequately developed to the extent that import is unnecessary will be prepared and periodically updated, in such areas no import of technology would normally be permitted; and the onus will be on the seeker of foreign technology, be it industry or a user Ministry, to demonstrate to the satisfaction of the approval authority that import is necessary.

Technology assessment systems will be reviewed. A technology assessment mechanism consisting of competent groups will render advice in all cases of technology import relating to highly sophisticated technology, large investments and national security. Aspects of employment energy, efficiency and environment will be kept in view.

The basic principles governing the acquisition of technology will be:

a) Import of technology, and foreign investment in this regard, will continue to be permitted only on a selective basis where; need has been established; technology does not exist within the country; the time taken to generate the technology indigenously would delay the achievement of development targets.
b) Government may, from time to time, identify and notify such areas of high national priority, in respect of which procedures would be simplified further to ensure timely acquisition of the required technology.
c) There shall be firm commitment for absorption adaptation and subsequent development of imported know-how through adequate investment in Research and Development to which importers of technology will be expected to contribute.

Unpackaging

Technology to fulfill a particular need consists of many components. It is necessary to develop capability to break down the total package of technology required for a purpose into components, some of which may be readily available or could be indigenously developed and others that will need to be imported. Norms and guidelines for such unpackaging will be evolved.

Absorption of Technology

There will be a commitment to ensure an adequate scale of investment in R&D for the absorption adaptation and wherever possible, improvement on and generation of new technology, making fullest use of overall national capabilities. Only thus can self-reliance be ensured and a technology generation process established firmly. Appropriate mechanism will be evolved at the stage of technology assessment to ensure the absorption of imported technology.

Technological Information

The availability of an efficient system of collection and analysis of relevant technological information including cost and other economic aspects, is a prerequisite for the appropriate choice of technologies. This will considerably enhance the possibility of obtaining favourable terms and conditions in acquisition of technology. Such a technology information base will be established.

Technology Transfer

Diffusion

Specific efforts need to be made for the diffusion of technology in use to all beneficiaries who can employ them optimally. Appropriate measures shall be evolved to facilitate technological diffusion, including horizontal transfer; technological support of ancillaries from large units; technology inputs to small units; and upgradation of traditional skills and capabilities.

International Competitiveness and Technology Exports

It is necessary to maintain international competitiveness in products, services and technologies that have export potential.

Conditions for the marketing of indigenous technology and of products base on it will be improved. It is important in all such cases to conform to the highest international standards.

Technical Cooperation among Developing Countries

A concerted effort will be made to participate fully in technical cooperation among developing countries. Encouragement will be provided for participation in technology development programmes with other developing countries which can contribute to mutual national development.

Protection: Legislative Framework

Development of technology calls for large investments and often involves considerable risk. Encouragement will be given to obtaining necessary protection in all cases of indigenous technology development. A mechanism will be set up to ensure that national interests arising from the generating of technology are fully protected internationally in terms of industrial property rights.

Implementation

The success of the Technology Policy and the speed with which the various facets of the policy are implemented will depend to a considerable extent on a system of efficient monitoring review and guidance and the scheme of incentives and disincentives.

Government will evolve instruments for the implementation of this Technology Policy and spell out in detail guidelines for Ministries and agencies of Government as well as for industries and entrepreneurs.

Success in implementation demands a conscious integrated approach covering technology assessment, development, acquisition, absorption, utilization and diffusion, and connected aspects of financing, based on overall national interests, priorities and the attainment of the most challenging technological goals.

Above, all the entire population must be imbued with self-confidence and pride in national capacity.

Indian Science and Technology must unlock the creative potential of our people and help in building the India of our dreams.

Annexure 3.3

Science and Technology Policy 2003

Science and technology have profoundly influenced the course of human civilization. Science has provided us remarkable insights into the world we live in. The scientific revolutions of the 20th century have led to many technologies, which promise to herald wholly new eras in many fields. As we stand today at the beginning of a new century, we have to ensure fullest use of these developments for the well being of our people.

Science and technology have been an integral part of Indian civilization and culture over the past several millennia. Few are aware that India was the fountainhead of important foundational scientific developments and approaches. These cover many great scientific discoveries and technological achievements in mathematics, astronomy, architecture, chemistry, metallurgy, medicine, natural philosophy and other areas. A great deal of this travelled outwards from India. Equally, India also assimilated scientific ideas and techniques from elsewhere, with open-mindedness and a rational attitude characteristic of a scientific ethos. India's traditions have been founded on the principles of universal harmony, respect for all creation and an integrated holistic approach. This background is likely to provide valuable insights for future scientific advances. During the century prior to Independence, there was an awakening of modern science in India through the efforts of a number of outstanding scientists. They were responsible for great scientific advances of the highest international caliber.

In the half century since Independence, India has been committed to the task of promoting the spread of science. The key role of technology as an important element of national development is also well recognised. The Scientific Policy Resolution of 1958 and the Technology Policy Statement of 1983 enunciated the principles on which the growth of science and technology in India has been based. Over the past several decades, these policies have emphasized self-reliance, as also sustainable and equitable development. They embody a vision and strategy that are applicable today, and would continue to inspire us in our endeavours.

With the encouragement and support that has been provided, there is today a sound infrastructural base for science and technology. These include research laboratories, higher educational institutions and highly skilled human resource. Indian capabilities in science and technology cover an impressive range of diverse disciplines, areas of competence and of applications. India's strength in basic research is recognized internationally. Successes in agriculture, health care, chemicals and pharmaceuticals, nuclear energy, astronomy and astrophysics, space technology and applications, defense research, biotechnology, electronics, information technology and oceanography are widely acknowledged. Major national achievements include very significant increase in food production, eradication or control of several diseases and increased life expectancy of our citizens.

While these developments have been highly satisfying, one is also aware of the dramatic changes that have taken place, and continue to do so, in the practice of science, in technology development, and their relationships with, and impact on, society. Particularly striking is the rapidity with which science and technology is moving ahead. *Science is becoming increasingly inter- and multi-disciplinary, and calls for multi-institutional and, in several cases, multi-country participation.* Major experimental facilities, even in several areas of basic research, require very large material, human and intellectual resources. Science and technology have become so closely intertwined, and so reinforce each other that, to be effective, any policy needs to view them together. The continuing revolutions in the field of information and communication technology have had profound impact on the manner and speed with which scientific information becomes available, and scientific interactions take place.

Science and technology have had unprecedented impact on economic growth and social development. Knowledge has become a source of economic might and power. This has led to increased restrictions on sharing of knowledge, to new norms of intellectual property rights, and to global trade and technology control regimes. Scientific and technological developments today also have deep ethical, legal and social implications. There are deep concerns in society about these. The ongoing globalization and the intensely competitive environment have a significant impact on the production and services sectors. Because of all this, our science and technology system has to be infused with new vitality if it is to play a decisive and beneficial role in advancing the well being of all sections of our society. The nation continues to be firm in its resolve to support science and technology in all its facets. It recognizes its central role in raising the quality of life of the people of the country, particularly of the disadvantaged sections of society, in creating wealth for all, in making India globally competitive, in utilizing natural resources in a sustainable manner, in protecting the environment and ensuring national security.

Policy Objective

Recognizing the changing context of the scientific enterprise, and to meet present national needs in the new era of globalisation, Government enunciates the following objectives of its Science and Technology Policy:

- To ensure that the message of science reaches every citizen of India, man and woman, young and old, so that we advance scientific temper, emerge as a progressive and enlightened society, and make it possible for all our people to participate fully in the development of science and technology and its application for human welfare. Indeed, science and technology will be fully integrated with all spheres of national activity.
- To ensure food, agricultural, nutritional, environmental, water, health and energy security of the people on a sustainable basis.
- To mount a direct and sustained effort on the alleviation of poverty, enhancing livelihood security, removal of hunger and malnutrition, reduction of drudgery and regional imbalances, both rural and urban, and generation of employment, by using scientific and technological capabilities along with our traditional knowledge pool. This will call for the generation and screening of all relevant technologies, their widespread dissemination through networking and support for the vast unorganized sector of our economy.
- To vigorously foster scientific research in universities and other academic, scientific and engineering institutions; and attract the brightest young persons to careers in science and technology, by conveying a sense of excitement concerning the advancing frontiers, and by creating suitable employment opportunities for them. Also to build and maintain centers of excellence, which will raise the level of work in selected areas to the highest international standards.

- To promote the empowerment of women in all science and technology activities and ensure their full and equal participation.
- To provide necessary autonomy and freedom of functioning for all academic and R&D institutions so that an ambience for truly creative work is encouraged, while ensuring at the same time that the science and technology enterprise in the country is fully committed to its social responsibilities and commitments.
- To use the full potential of modern science and technology to protect, preserve, evaluate, update, add value to, and utilize the extensive knowledge acquired over the long civilizational experience of India.
- To accomplish national strategic and security-related objectives, by using the latest advances in science and technology.
- To encourage research and innovation in areas of relevance for the economy and society, particularly by promoting close and productive interaction between private and public institutions in science and technology. Sectors such as agriculture (particularly soil and water management, human and animal nutrition, fisheries), water, health, education, industry, energy including renewable energy, communication and transportation would be accorded highest priority. Key leverage technologies such as information technology, biotechnology and materials science and technology would be given special importance.
- To substantially strengthen enabling mechanisms that relate to technology development, evaluation, absorption and upgradation from concept to utilization.
- To establish an Intellectual Property Rights (IPR) regime which maximizes the incentives for the generation and protection of intellectual property by all types of inventors. The regime would also provide a strong, supportive and comprehensive policy environment for speedy and effective domestic commercialization of such inventions so as to be maximal in the public interest.
- To ensure, in an era in which information is key to the development of science and technology, that all efforts are made to have high-speed access to information, both in quality and quantity, at affordable costs; and also create digitized, valid and usable content of Indian origin.

- To encourage research and application for forecasting, prevention and mitigation of natural hazards, particularly, floods, cyclones, earthquakes, drought and landslides.
- To promote international science and technology cooperation towards achieving the goals of national development and security, and make it a key element of our international relations.
- To integrate scientific knowledge with insights from other disciplines, and ensure fullest involvement of scientists and technologists in national governance so that the spirit and methods of scientific enquiry permeate deeply into all areas of public policy making.

It is recognized that these objectives will be best realized by a dynamic and flexible Science and Technology Policy, which can readily adapt to the rapidly changing world order. This Policy, reiterates India's commitment to participate as an equal and vigorous global player in generating and harnessing advances in science and technology for the benefit of all humankind.

Strategy and Implementation Plan

Keeping in view these broad objectives, it is essential to spell out an implementation strategy that will enable identification of specific plans, programmes and projects, with clearly defined tasks, estimates of necessary resources, and time targets. Some of the key elements of the implementation strategy will be as follows: -

1. *Science and Technology Governance and Investments:* Suitable mechanism will be evolved by which independent inputs on science and technology policy and planning are obtained on a continuous basis from a wide cross section of scientists and technologists. It will utilize the academies and specialized professional bodies for this purpose. These inputs will form an integral part of the planning and implementation of all programmes relating to science and technology, as also in government decision-making and formulation of policies in socio-economic sectors.

A greater integration of the programmes in socio-economic sectors with R&D activities will go a long way in ensuring a wider, more visible and tangible impact. This will call for a certain percentage of the overall allocation of each of the socio-economic ministries to be devoted for relevant programmes and activities in science and technology. The States will also be encouraged and assisted in the use of science and technology for developmental purposes through mechanisms set up for this, and in establishing linkages with national institutions for solving their regional and locale-specific problems.

A concerted strategy is necessary to infuse a new sense of dynamism in our science and technology institutions. The science departments, agencies and other academic institutions, including universities i.e. the science and technology system as a whole, would be substantially strengthened, given full autonomy and flexibility, and de-bureaucratized. Mechanisms will be established to review on a continuous basis the academic and administrative structures and procedures in the science and technology system at all levels, so that reforms could be effected to meet the challenges of the changing needs.

It will be ensured that all highly science-based Ministries/ Departments of Government are run by scientists and technologists. All the major socio-economic Ministries will have high-level scientific advisory mechanisms. Government will ensure continued existence of an Apex S&T Advisory Body, which will assist in formulating and implementing various programmes and policies. It will have appropriate representation of industry leaders, leading scientists and technologists and various scientific departments.

Government will make necessary budgetary commitments for higher education and science and technology. It will, through its own resources and also through contribution by industry, raise the level of investment to at least 2% of GDP on science and technology by the end of the Tenth Plan. For this, it is essential for industry to steeply increase its investments in R&D. This will enable it to be competitive, achieve greater self-reliance and self-confidence, and fulfill national goals.

2. *Optimal Utilization of Existing Infrastructure and Competence:* Science and technology is advancing at a very fast pace, and obsolescence of physical infrastructure, as also of skills and competence, take place rapidly. Steps will be taken to network the existing infrastructure, investments and intellectual strengths, wherever they exist, to achieve effective and optimal utilization, and constantly upgrade them to meet changing needs.

3. *Strengthening of the Infrastructure for Science and Technology in Academic Institutions:* A major initiative to modernize the infrastructure for science and engineering in academic institutions will be undertaken. It will be ensured that all middle and high schools, vocational and other colleges will have appropriately sized science laboratories. Science, engineering and medical departments in academic institutions and universities and colleges will be selected for special support to raise the standard of teaching and research. To begin with, a significant number of academic institutions, specially the universities, as also engineering and medical institutions, would be selected for this support to make an impact. Flexible mechanisms for induction of new faculty in key areas of science would be developed. Constancy of support and attention will be ensured over at least a ten-year period.

4. *New Funding Mechanisms for Basic Research:* The setting up of more efficient funding mechanisms will be examined, either by creating new structures or by strengthening or restructuring the existing ones, for promotion of basic research in science, medical and engineering institutions. In particular, administrative and financial procedures will be simplified to permit efficient operation of research programmes in diverse institutions across the country.

Creation of world class facilities in carefully selected and nationally relevant fields will be undertaken, to enhance our international competitiveness in areas where we have strengths, opportunities or natural advantages. Indigenous expertise will be used to the maximum extent possible. This would help in nurturing high quality talent and expertise in experimental science and engineering.

5. *Human Resource Development:* The number of scientists and technologists, while being large in absolute numbers, is not commensurate with the requirements in quality and when measured on a per capita basis. The demand is bound to increase in the coming years with more intensive activities involving science and technology. There is need to progressively increase the rate of generation of high quality skilled human resource at all levels. This process would naturally entail reversing the present flow of talent away from science, by initiating new and innovative schemes to attract and nurture young talent with an aptitude for research, and by providing assured career opportunities in academia, industry, Government or other sectors. *In*

order to encourage quality and productivity in science and technology, mobility of scientists and technologists between industry, academic institutions and research laboratories will be ensured.

For building up the human resource base in relevant areas, the agencies and departments concerned with science and technology will make available substantial funding from their allocation. Flexible mechanisms will be put in place in academic and research institutions to enable researchers to change fields and bring new inputs into traditional disciplines, and also to develop inter-disciplinary areas. There will be emphasis on a continuing process of retraining and reskilling to keep pace with the rapid advances taking place. Wherever considered necessary, training abroad will be resorted to, so as to build up a skilled base rapidly.

Women constitute almost half the population of the country. They must be provided significantly greater opportunities for higher education and skills that are needed to take up R&D as a career. For this, new procedures, and flexibility in rules and regulations, will be introduced to meet their special needs.

New mechanisms would be instituted to facilitate the return of scientists and technologists of Indian origin to India, as also their networking, to contribute to Indian science and technology. Schemes for continuing education and training of university and college teachers in contemporary research techniques and in emerging areas of science will be strengthened and new innovative programmes started. It will also be ensured that higher education is available to the widest possible section of creative students, transcending social and economic barriers.

6. *Technology Developments, Transfer and Diffusion:* A strong base of science and engineering research provides a crucial foundation for a vibrant programme of technology development. Priority will be placed on the development of technologies, which address the basic needs of the population; make Indian industries — small, medium or large — globally competitive; make the country economically strong; and address the security concerns of the nation. Special emphasis will be placed on equity in development, so that the benefits of technological growth reach the majority of the population, particularly the disadvantaged sections, leading to an improved quality of life for every citizen of the country. These aspects require technology foresight,

which involves not only forecasting and assessment of technologies but also their social, economic and environmental consequences.

The growth rate in productivity of the Indian economy has been below its true potential, and the contribution to it of technological factors is inadequate. Similarly, Indian exports today derive their comparative advantage through resource and labour rather than through the power of technological innovation. The transformation of new ideas into commercial successes is of vital importance to the nation's ability to achieve high economic growth and global competitiveness. Accordingly, special emphasis will be given not only to R&D and the technological factors of innovation, but also to the other equally important social, institutional and market factors needed for adoption, diffusion and transfer of innovation to the productive sectors.

Intensive efforts will be launched to develop innovative technologies of a breakthrough nature; and to increase our share of high-tech products. Aggressive international bench-marking will be carried out. Simultaneously, efforts will be made to strengthen traditional industry so as to meet the new requirements of competition through the use of appropriate science and technology. This industry is particularly important as it provides employment at lower per capita investment, involves low energy inputs, and carries with it unique civilizational traditions and culture. Value addition, and creation of wealth through reassessment, redistribution and repositioning of our intellectual, capital and material resource will be achieved through effective use of science and technology.

Deriving value from technology-led exports and export of technologies will be facilitated through new policy initiatives, incentives and legislation. This will include intensive networking of capabilities and facilities within the country. Rigid Quality Standards, and Accreditation of testing and calibration laboratories according to international requirements, will be given an enhanced push to enable Indian industry to avoid non-tariff barriers in global trade.

A comprehensive and well-orchestrated programme relating to education, R&D and training in all aspects of technology management will be launched. To begin with, Indian Institutes of Management (IIMs), Indian Institutes of Technology (IITs) and other selected institutions will be encouraged to initiate these programmes.

7. *Promotion of Innovation:* Innovation will be supported in all its aspects. A comprehensive national system of innovation will be created covering science and technology as also legal, financial and other related aspects. There is need to change the ways in which society and economy performs, if innovation has to fructify.

8. *Industry and Scientific R&D:* Every effort will be made to achieve synergy between industry and scientific research. Autonomous Technology Transfer Organizations will be created as associate organizations of universities and national laboratories to facilitate transfer of the know-how generated to industry. Increased encouragement will be given, and flexible mechanisms will be evolved to help, scientists and technologists to transfer the know-how generated by them to the industry and be a partner in receiving the financial returns. Industry will be encouraged to financially adopt or support educational and research institutions, fund courses of interest to them, create professional chairs etc. to help direct S&T endeavours towards tangible industrial goals.

There has to be increased investments by industry in R&D in its own interest to achieve global competitiveness to be efficient and relevant. Efforts by industry to carry out R&D, either in-house or through outsourcing, will be supported by fiscal and other measures. To increase their investments in R&D, innovative mechanisms will be evolved.

9. *Indigenous Resources and Traditional Knowledge:* Indigenous knowledge, based on our long and rich tradition, would be further developed and harnessed for the purpose of wealth and employment generation. Innovative systems to document, protect, evaluate and to learn from India's rich heritage of traditional knowledge of the natural resources of land, water and bio-diversity will be strengthened and enlarged. Development of technologies that add value to India's indigenous resources and which provide holistic and optimal solutions that are suited to Indian social-cultural-economic ethos will be developed.A concerted plan to intensify research on traditional systems of medicine, so as to contribute to fundamental advances in health care, and leading to commercialisation of effective products will be undertaken; appropriate norms of validation and standardization will be enforced. A purposeful programme to enhance the Indian share of the global herbal product market will be initiated.

10. *Technologies for Mitigation and Management of Natural Hazards:* Science and technology has an important role in any general strategy to address the problems of mitigation and management of the impacts of natural hazards. A concerted action plan to enhance predictive capabilities and preparedness for meeting emergencies arising from floods, cyclones, earthquakes, drought, landslides and avalanches will be drawn up. Measures will be undertaken to promote research on natural phenomena that lead to disasters and human activities that aggravate them. This will be with a view to developing practical technological solutions for pre-disaster preparedness, and mitigation and management of post- disaster situations.

11. *Generation and Management of Intellectual Property:* Intellectual Property Rights (IPR), have to be viewed, not as a self-contained and distinct domain, but rather as an effective policy instrument that would be relevant to wide ranging socio-economic, technological and political concepts. The generation and fullest protection of competitive intellectual property from Indian R&D programmes will be encouraged and promoted.

The process of globalisation is leading to situations where the collective knowledge of societies normally used for common good is converted to proprietary knowledge for commercial profit of a few. Action will be taken to protect our indigenous knowledge systems, primarily through national policies, supplemented by supportive international action. For this purpose, IPR systems which specially protect scientific discoveries and technological innovations arising out of such traditional knowledge will be designed and effectively implemented.

Our legislation with regard to Patents, Copyrights and other forms of Intellectual Property will ensure that maximum incentives are provided for individual inventors, and to our scientific and technological community, to undertake large scale and rapid commercialization, at home and abroad. The development of skills and competence to manage IPR and leveraging its influence will be given a major thrust. This is an area calling for significant technological insights and legal expertise and will be handled differently from the present, and with high priority.

12. *Public Awareness of Science and Technology:* There is growing need to enhance public awareness of the importance of science and

technology in everyday life, and the directions where science and technology is taking us. People must be able to consider the implications of emerging science and technology options in areas which impinge directly upon their lives, including the ethical and moral, legal, social and economic aspects. In recent years, advances in biotechnology and information technology have dramatically increased public interest in technology options in wide ranging areas. Scientific work and policies arising from these have to be highly transparent and widely understood.

Support for wide dissemination of scientific knowledge, through the support of science museums, planetaria, botanical gardens and the like, will be enhanced. Every effort will be made to convey to the young the excitement in scientific and technological advances and to instill scientific temper in the population at large. Special support will be provided for programmes that seek to popularize and promote science and technology in all parts of the country. Programmes will also be developed to promote learning and dissemination of science through the various national languages, to enable effective science communication at all levels. A closer interaction of those involved in the natural sciences and technology, social sciences, humanities and other scholarly pursuits will be facilitated to bring about mutual reinforcement, added value and impact.

13. *International Science and Technology Cooperation:* Scientific research and technology development can benefit greatly by international cooperation and collaboration. Common goals can be effectively addressed by pooling both material and intellectual resources. International collaborative programmes, especially those contributing directly to our scientific development and security objectives, will be encouraged between academic institutions and national laboratories in India and their counterparts in all parts of the world, including participation in mega science projects as equal partners. Special emphasis will be placed on collaborations with other developing countries, and particularly neighbouring countries, with whom India shares many common problems. International collaboration in science and technology would be fully used to further national interests as an important component of foreign policy initiatives.

14. *Fiscal Measures:* Innovative fiscal measures are critical to ensure successful implementation of the policy objectives. New methods are required for incentivising R&D activities, particularly in industry. New strategies have to be formulated for attracting higher levels of public and private investments in scientific and technological development. A series of both tax and non-tax fiscal instruments have to be evolved to ensure a leap-frogging process of development. The formulation of a focused strategy and the designing of new methods and instruments requires inputs from economists, financial experts and management experts and scientists. For this purpose, the apex S&T advisory body will constitute a dedicated task-force to suggest appropriate fiscal measures to subserve the policy objectives.

15. *Monitoring:* Effective, expeditious, transparent and science-based monitoring and reviewing mechanisms will be significantly strengthened, and wherever not available will be put in place. It will be ensured that the scientific community is involved in, and responsible for, smooth and speedy implementation.

16. *The New Vision:* To build a new and resurgent India that continues to maintain its strong democratic and spiritual traditions, that remains secure not only militarily but also socially and economically, it is important to draw on the many unique civilizational qualities that define the inner strength of India; this has been intrinsically based on an integrated and holistic view of nature and of life. The Science and Technology Policy 2003 will be implemented so as to be in harmony with our world view of the larger human family all around. It will ensure that science and technology truly uplifts the Indian people and indeed all of humanity.

Chapter 4

Dissemination of Research

Both the Scientific Policy Resolution of 1958 and the Technology Statement of 1983 recognized the essentiality of acquisition and dissemination of research. The Resolution thus aimed for (a) encouraging individual initiative for acquisition and dissemination of knowledge in an atmosphere of academic freedom and (b) securing for the people benefits that can accrue from acquisition and application of scientific knowledge. It may be gainsaid that both these issues have a bearing on '*intellectual property rights*'. Somehow, its significance was not sufficiently realized at the time when the science and technology policy resolutions were being formulated. As it has been observed, 'producing manpower' rather than *participation in technology revolutions* became the prime objective of the Indian educational system (Bhattacharya, 1994). In contrast, 'one can trace well-laid out "Intellectual Property Rights" policies in most universities in advanced countries *as ownership of the right,* its manifestation and obligation is a matter of routine professional concern (Bhattacharya, 1994).

To be fair enough, the aims and objectives of the Government were firstly to inculcate a scientific temper in a traditional society. Acquisition of knowledge called for creating a network of libraries and laboratories. Similarly, dissemination of knowledge acquired in laboratories required extension services. There has, nonetheless, been the usual criticism of obsolescence in knowledge in college and university education. It was noted that research activity and teaching got segregated *with the establishment of stand-alone research institutes. Whereas* textbook teaching became the private responsibility of the universities, R&D supposedly belonged to the government owned laboratories and/or industries (Bhattacharya, 1994). The National Policy on Education (1986) and the Science and Technology Policy Statement (2003) made some effort to correct the situation.

The Science and Technology Statement (2003) attaches a good deal of significance to Intellectual Property Rights (Annexure 3.3).

4.1 Extension Services

The National Commission on Agriculture (1976), Government of India, defined 'extension service' as *out of school education*. Generally, it refers to *transfer of technology* from laboratories/research institutions to direct producers/beneficiaries. 'Independent India acknowledged the relevance of extension quite early; a decade earlier than organized attempts to strengthen agricultural research was initiated in the country'.(Sulaiman 2003) Agricultural extension was, however, provided to the farmers until the 1960's at the initiative of the State Departments of Agriculture. The Indian Council of Agricultural Research (ICAR) under the Ministry of Agriculture & Co-operation, Government of India began to get involved in extension activities with the introduction of the *National Demonstration Programme* in 1966. The 'lab to land' programme was also largely based on demonstrations.

ICAR's involvement further increased with the introduction of Krishi Vigyan Kendra (KVK). The KVKs are the field level *farm science centres* for vocational training and technology transfer. "Teaching by doing" and "learning by doing" and "work experience" are the main principles of proxiology for imparting skill training. The first KVK was established in Pondicherry in 1974. Since then, this network has expanded and numbers 314 today. So far, the KVKs have been supported by the ICAR. It is intended now to broad base the same with the involvement of the corporate sector/NGOs, farmer's organizations, cooperatives and agri-clinics. The KVKs may also function as nodal agencies for quality certification including organic products, bio-fertilizers and bio-pesticides.

Under the Community Development Programme initiated in the early 1950's, the Village Level Workers (VLWs) and the Agricultural Extension Officers (AEOs) were the important links between the scientific community and the farmers. The subsequent involvement of the educated in rural India, including the elected representatives of Panchayati Raj, made extension a success and heralded *'the green revolution'* in Indian agriculture. Now onwards the extension service in agriculture would be mainly provided by (5000) 'agri-clinics' managed by agricultural graduates. The agricultural graduates are given training in extension at the earmarked universities/institutions. While banks would mainly finance the agri-clinics by way of loans, the Government of India would provide 25% of the cost as subsidy. The average cost for each individual venture is estimated to be Rs.5 lakh (which could go to Rs.10 lakh in special cases). If two or more individuals undertook a joint/group venture, the combined project could be considered for support under the scheme *in terms of multiples of Rs.10 lakh*. The cost of such venture could go up to a maximum of Rs.50 lakh for a group of 5 or more applicants (Sulaiman, 2003) .

The nation wide programme of Integrated Child Development Scheme (ICDS) is yet another example of extension service. It is run with the help of Anganwadi Workers (AWWs) who are paid an honorarium for their services. The Anganwadi centres are also visited by the Auxiliary Nurse Midwives (ANMs) appointed on a contractual basis by the Department of Health of State Governments. The ICDS caters to pre-school children of six years and expectant and nursing mothers, with a package of services like health check-up, immunization, referral services, supplementary nutrition, pre-school education and health and nutrition education. The AWWs have been providing the (extension) services by way of giving instructions in health care, such as treatment of acute respiratory infections, case of newborn and management of neo-natal infections. There is a need to develop similar concepts, such as, the agri-clinics to upgrade the ICDS scheme.

4.2 Training Policy for the Civil Services

The National Training Policy (1998) likewise viewed training as the most cost effective method of improving competence of the officers and staff of the Government. It accordingly recommended for adoption of 5% of the budget to be spent on trainings as the norm for all Government departments. The Government of India is said to have brought down this figure to 1% of the annual budget of each department. Policy towards training has been further liberalized and middle level officers are being encouraged to go for training programmes at institutions of excellence abroad, through partial funding.

According to the NTP, moreover, 'since the civil services system has been formed for the achievement of constitutionally defined goals of the state, *training designs should address themselves to the training objectives and concerns for all strata of civil service in operational terms'*. Since training has to be organized for the entire range of civil service system, covering all levels from the lowest to the highest. It broadly classified the civil servants into three categories/levels, namely;

a) the lowest level functionaries, at the cutting edge with direct interface with public who are mostly members of the Group 'D' and Group 'C' services.
b) The supervisory levels and middle management level who are mostly members of the Group 'B' services;
c) Group 'A' services and All India Service comprising the top administration/management levels.

It further recommended that 'training programmes at the cutting edge level should put emphasis on functional skill and attitudinal orientation. For the middle level, training should contribute to enhancement of professional knowledge, understanding and skill, and at the same time promote widened professional out look so as to prepare for further tasks and responsibilities. For the third level, besides offering stimuli for expansion of the mental horizon and attainment of professional excellence *training should also sharpen perceptions in relation to multi-dimensional linkages in policy formulation*, development capabilities in a changing socio-political and technological context, dynamism and innovation tempered with pragmatism'.

4.3 Refresher Courses for University Education

A beginning was made to address the criticism of obsolescence in college and university education due to isolation from research activity, through introducing refresher and orientation courses. This was based on the recommendations made as part of the National Policy on Education (1986). Accordingly, the UGC formulated a scheme of Academic Staff College (ASCs) in 1987-88 for organizing *orientation programmes* for newly appointed lecturers and *refresher courses* for in-service teachers. The scheme of ASCs was subsequently reviewed in 1991 and following suggestions were made to further improve the academic standards.

1. The recommendations of the UGC Review Committee on Academic Staff Colleges should be implemented at the earliest.

2. Before considering further expansion of Academic Staff Colleges, UGC and the universities concerned should take measures for consolidation and strengthening of the existing ASCs;
3. Efforts should be made to initiate training and orientation of teachers in engineering and technical subjects;
4. UGC should undertake a five yearly review of the performance of the ASCs;
5. Keeping in view the popularity of IGNOU's Diploma in Distance Education and Higher Education, suitable distance education programmes should be designed for upgrading knowledge and skills of teachers in institutions of higher education.
6. Efforts of UGC and IGNOU for training of teachers should be integrated and coordinated. For instance ASCs could organize contact programmes for those teachers who are enrolled in IGNOU's Diplomas in Distance and Higher Education.

As already mentioned one of the important recommendations of the NPE was that of encouraging public and private sector enterprises to set up at least some of their R&D activities on the campuses of higher education. It also advocated for encouraging public sector research organizations, such as, the CSIR and the DRDO to implement some of their R&D activities through universities.

4.4 Intellectual Property Rights

Participation in technology revolution through acquiring 'ownership right' is often a big motivating factor in the pursuit of research. Ownership right or 'appropriability' has, at the same time, the negative aspect of restricting dissemination of research and diffusion of knowledge. Intellectual Property Rights (IPR) usually refer to (i) patent right, (ii) copyright, (iii) trade and service mark (iv) tangible research property, (v) geographical indications and (vi) mask for integrated circuits and semi-conductors. Patenting is defended on the ground of recovering the cost incurred on the development of an innovation. Patent is a device or a way to reward invention where it has commercial potential. The reward is attributed both to the inventor and the institution (Bhattacharya, 1994). Copyright may be in regard to a publication, a performer's right or satellite broadcasting. The copyright, for a publication generally vests with the authors and they benefit from the payment of royalty. However, if the study is a sponsored one, the copyright might go to the sponsor/institute; although the sponsors may not necessarily insist on the copyright.

A trade or service mark is defined as any sign or a combination of signs capable of distinguishing the goods or services of one undertaking from those of other undertakings. They are related to products, events, activities, or services in the name of the trade/service/institute. Tangible research property, similarly, refers to research products which are of definite, concrete and expressed in material form, such as, engineering prototypes, engineering drawings, biological or chemical products or micro-circuit chips/computer software. 'Often tangible property rights may have association with intangible property rights which may be either copyrighted and/or patented' (Bhattacharya, 1994). Under the WTO Agreement, all agreeing parties/member countries are expected to enact laws to prevent use of any means in the designation or presentation of a good that indicates or suggests that the good in question originates in a geographical area other than the true place of origin ('geographical indications'). Geographical indications, thus, form part of IPR.

Masks are patterns (geometrical) on plates that contain design information of microcircuits. The US universities are protecting masks relating to IPR through copyright. In the Indian context, it is noticed that a fair amount of research is sponsored. Most of the research grants are from the union or the state governments and the IPR vests with the Government. However, the sponsoring agency hardly expects any commercial return and most of the projects are open-ended and meant for competence building.

It is only now that ownership rights over 'know-how'/ 'know-why' is being mentioned and its commercial significance acknowledged (Bhattacharya, 1994). The Science and Technology Statement (2003), thus, makes the following policy pronouncement in regard to IPR, namely, 'The generation and fullest protection of competitive intellectual property from Indian R&D programmes will be encouraged and promoted… Our legislation with regard to patents, copyrights and other forms of Intellectual Property will ensure that maximum incentives are provided to individual inventors and to our scientific and technological community for undertaking large scale and rapid commercialization, at home and abroad. The development of skills and competence to manage IPR and leveraging its influence will be given a major thrust'. Accordingly, it been suggested that *one credit course on IPR* be introduced in all universities/institutions of higher learning for providing the necessary information to research students.

4.5 Libraries and Research

As Chairman of the University Education Commission (1948-49) Dr. Sarawpalli Radhakrishnan (later on President of India) observed, 'Teaching is a cooperative enterprise. Teachers must have the necessary tools for teaching purposes in the shape of libraries and laboratories and also the right type of students. The library is the heart of all the university work, directly so, as regards its research work, and indirectly as regards its educational work, which derives its life from research work. He further opined, 'Scientific research needs a library as well as its laboratories while for humanities research the library is both library and laboratory in one' (Taher, 2001).

In recent history, the Western Library Movement has had enthusiasts who contributed voluntarily all that was possible by them in the form of cash, kind and personal services. The Englishmen (In India) who came as officers or Session Judges to Bombay and Madras similarly are credited to have set up small libraries in the localities where they were stationed with the help of the elite of the town. The governments to-day play a prominent role in promotion of libraries through enactments of relevant laws, development and maintenance of libraries and regularization of such establishments, especially when private funding is insufficient to meet the requirements.

While libraries are basic to research, the following observations in regard to the status of libraries in India is indeed alarming, namely, 'it has always been assured that the academics in an institution of higher learning will have access through journals and books to the whole fund of knowledge generated in the world and will be current to within a few months of actual publication. This assumption is completely invalid in the circumstances that prevail in our country'. A large majority of our 4 million higher education students and 250,000 teachers in the universities and colleges can read about what is new only in old review journals and popular magazines and that too in a form which is not very useful to an active researcher; this will not give any one a feeling of being a participant in the enterprise of creating new knowledge' (Yash Pal, 1988).

There are indeed competing demands from various sectors for the limited resources available with the Government. An affordable way to address the plight of libraries was found by way of setting up an Information and Library Network (INFLIBNET), which would inter-connect the people and library resources of all universities/ institutes of higher education and research laboratories. It also aimed to keep a record of all the Ph.D thesis produced in the Indian Universities and make them available to any researcher, making request for the same from any part of the country. The scheme was launched in 1991, but it has yet to make an impact on the state of libraries in India. The Department of Culture, Ministry of Human Resource Development had earlier set up an Empowered Committee (November 1987) as a follow-up to the formulation of the National Policy in Library and Information System. The Committee addressed itself mainly to the needs of (i) Special Library System (ii)Academic Library System, (iii) Special Library & Information System and (iv) National Library and Bibliography Services.

In regard to the Library Services in India, the Report of the Working Group on Development of Information and Library Network (UGC 1988) made the following observations, "There would be around 2,000 to 2,500 special libraries in the country, in fields such as science and technology, medicine, agriculture, social sciences and arts and humanities. The special libraries have, on an average, 30,000 – 40,000 volumes of stock, add about 400 volumes per year, receive 300 periodicals, spend Rs.5.00 lakh each year on books and periodicals, have a professional staff strength of 5 to 15 and serve an active clientele in the range of 50 to 500, depending upon the size of the parent body. The country has national level libraries such as National Library, Kolkata, National Science Library (INSDOC), New Delhi, National Medical Library (DGHS), New Delhi and the IARI Library, New Delhi. The public library system, though weak in many states, has vast scope and potential and needs development".

Table 4.1 shows the plan expenditure incurred/ proposed to be incurred on major libraries by the Ministry of Culture, Government of India during the period 1999-2000 to 2004-05.

The Working Group (UGC,1988) further observed; 'in the post-independence period, there has been a gradual evolution of bibliographical information centers and systems. At the national level, there are INSDOC, DESIDOC, National Informatics Center, UGC Information Centres, BARC Library and Information Services Division, ICAR Agricultural Research Information Centre, National Documentation Centre for Health and Family Welfare, Small Enterprises National Documentation Centre (SENDOC) and NISSAT sectoral centers. In social science fields, there are NASSDOC, Centre for Rural Documentation in Hyderabad and a few others. In addition, there exist a number of subject specialized information centers/libraries attached to R&D and other institutions and industrial undertakings. Of them, the important ones may be in the range of 800-1000'.

Table 4.1 : Expenditure (Plan) on Libraries, Ministry of Culture, Government of India

(Rs.in '000)

S.No.	*Libraries*	*1999-2000 (A)*	*2000-2001 (A)*	*2001-2002 (A)*	*2002-2003 (RE)*	*2003-2004 (BE)*	*2004-2005*
1.	National Library, Kolkata	24500	29800	28700	35000	45000	600000
2.	Delhi Public Library	8000	8600	5600	10000	12000	30000
3.	Raja Ram Mohan Roy Library	5,68,00	72500	66600	92500	85000	100000
4.	Nehru Memorial Museum Library	20000	18400	14100	17500	25000	26000
5.	Other Libraries	23200	26000	26200	51000	56700	307800
6.	Central Reference Library, Kolkata	4500	—	—	—	—	—

B.E. = Budget Estimate, R.E. = Revised Estimate.

N.B. : Other libraries include Central Reference Library (Kolkatta), Central Library (Mumbai). Khuda Baksh Oriental Public Library (Patna), Rampur Raza Library (Thanjerun), Saraswati Mahal Library (Thanjevur) and Connemare Library (Chennai).

Chapter 5

Funding Windows for Research

Research may be conducted *in-house* in specialized institutions with a permanent staff or be conducted in universities on a short term basis. In-house research may often be preferable due to the capital intensive and long-term nature of research involving product development or new process creation. Barring such cases of highly capital-intensive projects, it is worthwhile organizing research through 'funding windows'. Extramural research or 'sponsored research' may go very well with the individual's interest and specialization. Since it is usually for a short duration, it is also more cost effective. Higher studies in universities are by nature closer to *going beyond the existing knowledge.* New questions arise and fresh doubts are raised; in the process, *there is greater creativity.* Going for further research to enhance knowledge, is, thus quite natural to these environments. This is true for both the teachers as well as for the students who may need financial assistance to pursue (basic) research.

Basic research thus enables the academia to be *in the lead* in regard to sciences and philosophy. The National Educational Policy (1986) quite categorically stated '*it is important that teachers get an opportunity to involve themselves in research activities'*. It further advocated for having a Dean or Board *associated with research* in educational institutions of higher learning and a percentage of grants-in-aid to be earmarked for research. The well-known universities the world over explicitly recognize the significance of research in their curriculum. Thus, the examination pattern in these universities at M.A./ M.Sc. level combines the *year-end* or *semester-end* examination (50%) with the (submission of) semester-long research (term) paper (50%). Some of the universities, abroad do not recognize the degree from India since most of the Indian universities fail to sufficiently reflect the research component. As such, while M.A. and M.Sc. degrees may not get recognized, PhD is recognized since that is evidence of research output.

The Central Government has been providing ad hoc grants for research in India since past several decades. Ad hoc grants are provided both for sponsored research as well as for 'wished for'/proposed research. Among the various ministries/departments of the Government of India, the roles of Ministry of Human Resource Development, Ministry of Science and Technology, Ministry of Health and Family Welfare, Ministry of Agriculture & Cooperation and Ministry of Environment and Forests are more significant. The Planning Commission has also been providing financial assistance to institutions and scholars, for research studies and for organizing workshops, conferences and consultations. Conferences and consultations help achieve both the objectives of (a) acquisition of research and (b) dissemination of research. It provides admirably the opportunity for exchange of ideas amongst scholars. Conferences and workshops when organized at the national or international level, compared to locally organized event, provide a wider choice of participants and speakers. This involves, quite obviously, a larger expenditure by way of transportation costs etc. Such meets need to be therefore organized more selectively and should ideally be backed up with publication of the presentations.

The paragraphs below introduce the main funding agencies under the different Ministries of the Government of India. The expenditure incurred on *sponsored research i.e., other than in-house research,* under each of these Ministries during the period 1998-99 to 2003-04 has also been briefly discussed. *The type of research programmes offered by the different implementing agencies, the eligibility criteria, the duration of research assistance etc has been indicated in Appendix I.* Various universities and philanthropic organizations, the world over, have been also *offering international fellowships/research assistance to encourage young students and scholars for pursuing research* and are the other examples of 'funding windows'. A representative list of *philanthropic organizations and universities abroad* offering fellowships to Indian students and scholars (along with the funding pattern) may be seen at *Appendix II.*

Admission to US universities for pursuing research (Ph.D) generally requires a minimum score in GRE tests sponsored by the Graduate Record Examination Board in the United States. Candidates are also required to qualify in TOEFL (Test of English as a Foreign

Language) examination. Higher Scores in GRE is helpful in winning a fellowship or 'aid' from the university. Another way to enable more students to undertake research in reputed institutions abroad in a cost effective manner, is through negotiating 'cultural exchange programmes' especially with countries where education is heavily subsidized. India has Cultural Exchange Programmes with a number of countries, including both advanced (e.g., Australia, France, Germany, Italy, Japan, Hungary, Finland, Sweden, Netherlands etc.) and developing (e.g., China, Sri Lanka, Thailand, Vietnam, Iran, Iraq, Kenya, Pakistan etc.) countries. One could obtain greater details about the various 'exchange programmes' from the Ministry of Culture (Government of India) and the Indian Council of Cultural Relations (ICCR), New Delhi and the embassies concerned with the country of study.

Ministry of Human Resources Development

The main agencies providing grants-in-aid to research institutes and individual scholars under the Ministry of Human Resources Development, Government of India are the University Grants Commission (UGC), the All India Council of Technical Education (AICTE), the Indian Council of Social Science Research (ICSSR), the Indian Institute of Advanced Studies (IIAS), the Indian Council of Historical Research (ICHR) and the Indian Council of Philosophical Research (ICPR). Altogether the Ministry disbursed more than six hundred crore as plan funds to these autonomous bodies during 2002-03, aimed mainly as grant for research studies. A similar sum was budgeted for research during the year 2003-04 (Table 5.1). The Ministry also provided grant-in-aid for organizing seminars and conferences in priority areas, such as, implementation of the National Education Policy (NEP) and issues relating to Copyrights and Intellectual Property Rights (IPR) etc. While the plan expenditure shown for the National Council of Educational Research and Training (NCERT) has generally been for in-house research for formulating major programmes in the field of school education, those shown for the Indira Gandhi National Open University (IGNOU) includes expenditure incurred on sponsored research, for preparing relevant material for different courses, with the help of consultants and university professors.

Table 5.1 : Expenditure on Sponsored Research –
Ministry of Human Resource Development

(Rs.in '000)

S.No.Scheme	*1998-1999 (RE)*	*1999-2000 (A)*	*2000-2001 (A)*	*2001-2002 (A)*	*2002-2003 (RE)*	*2003-2004 (BE)*
I. Grant-in-Aid for Research – (for fellowships/research studies)						
1. UGC	3303500	3837700	4070000	4397800	5080900	5167500
2. ICSSR	70000	74999	129000	152396	157500	175000
3. ICHR	7500	8994	17500	20663	25200	28000
4. ICPR	12500	11214	10145	21243	21600	24000
5. IAAS	9000	9741	24989	24613	27200	27500
6. NCERT	60000	100000	96700	83700	150500	140000
7. IGNOU	200000	380000	420000	525600	353000	670000
II. Grant-in-Aid for Seminars &Training						
1. Implementation of National Education Policy	10000	6000	5489	6641	7000	9000
2. Copyright matters	3000	1433	2063	7195	—	—
3. Intellectual Property Rights	2000	1031	2126	—	16000	28000
4. WTO	—	—	—	—	4000	10000
5. Promotion of Science Laboratories	—	—	—	—	—	—

BE = Budget Expenditure; RE = Revised Expenditure; A=Actual.
Source : Detailed Demand for Grants, Ministry of Human Resource Development, Government of India (Several Issues).

Somehow, *'promotion of science laboratories'* in educational institutions did not find favour by the Ministry in the form of any significant (separate) allocation of plan funds during these years. The data shown are, however, exclusive of the non-plan expenditure incurred on the salary of the staff of these autonomous bodies and for maintenance expenditure etc. The main areas of study, the type of fellowships offered and the eligibility criteria etc. under each of the apex bodies/agencies mentioned above may be seen at Appendix I.

Ministry of Science and Technology

The Ministry of Science and Technology, Government of India has been in the lead in promoting *research in newly emerging and front line areas of science and technology*. While it has established premier institutions for *in-house research* in areas ranging from basic sciences to weather forecasting, it has also been providing fellowships to students and scholars for pursuing independent extramural research. These grants-in-aid for research are not ad hoc in the sense, that these are provided on a regular basis; although the beneficiaries may be different individuals. The topic of research, similarly, may belong either to 'basic research' or 'applied research'. They are, however, generally of short duration. The Department of Science and Technology (DST), the Council of Scientific and Industrial Research (CSIR) and the Department of Bio-technology are the major funding agencies under the Ministry granting financial assistance, both to research scholars and research institutes. R&D support under the *mission mode*, *new millennium initiative*, *biotechnology research and industrial research* has been attracting greater attention of the policy and decision makers in recent years. The multidisciplinary approach to R&D programmes under the Science and Engineering Research Council (SERC) Scheme is increasingly acquiring the mission mode to research. The share of extramural research showed a steady increase over the past several years and went up from eight percent of the total research grant of CSIR in 2000-01 to fourteen per cent in 2003-04 (Table 5.2). The Ministry also provided financial support for organizing seminars and conferences and extended financial assistance to Indian scientists for going abroad for making presentations/participation at international symposiums.

Autonomous bodies like the Indian Academy of Science, (Bangalore), the National Academy of Science, (Allahabad), the Indian National Science Academy (New Delhi) and the Indian Science Congress Association, (Kolkata), have played very eminent roles in promotion of scientific research since independence and continue to receive grant-in-aid from the Government. (The Table 5.2 mentions only those schemes and institutions which have been important in the realm of extramural research under the Ministry). The type of fellowships offered by the CSIR, the main areas of study and the eligibility criteria may be seen at Appendix I.

Table 5.2 : Expenditure on Sponsored Research –
Ministry of Science and Technology (Rs/Lakh)

S.No. Scheme	*1998-1999 (A)*	*1999-2000 (A)*	*2000-2001 (A)*	*2001-2002 (RE)*	*2002-2003 (RE)*	*2003-2004 (BE)*
I. Grant-in-Aid for Research – (for fellowships/lab.facilities)						
1. CSIR (scholarships & fellowships)	–	–	450	550	500	700
2. CSIR (Total)	5784	5848	5110	–	5065	5000
3. CSIR (New Millenium Directive)	–	–	2500	4999	3000	3500
4. DST(SERC)	–	–	9967	100	–	–
5. DST (Mission Mode)			–	–	20700	21500
6. Deptt of Bio-Technology	823.5		2624.23	1099.98	–	–
7. Indian Academy of Science, Bangalore	80	80	80	80	80	80
8. National Academy of Science, Allahabad	30	–	39	95	165	165
9. Indian National Science Academy, New Delhi	300	300	350	350	350	310
10. Indian Science Congress Association, Calcutta	60	–	70	71	100	100
II. Grant-in-Aid for Seminars and Training						
1. Seminars, Symposia	185		202.23	194.80	200	200
2. Assistance to Indian Scientists going abroad	–	107.22	100	200	200	200
3. Vigyan Prasar	100	–	100	120	200	200

BE = Budget Expenditure; RE = Revised Expenditure. A= Actual
DST = Deptt. of Science and Technology;SERC=Science and Engineering Research Council;CSIR=Council of Scientific & Industrial Research
Source : Detailed Demand for Grants, Ministry of Science and Technology, Government of India (Several Issues).

Ministry of Health and Family Welfare

The Indian Council of Medical Research (ICMR) is the main agency under the Ministry of Health and Family Welfare for promoting bio-medical and health research. A great deal of in-house research is conducted in various premier institutions under the overall control of ICMR. The council also encourages extramural research conducted by independent scholars. Under the Cancer Research Programme financial assistance have been provided by way of gift of equipments to regional cancer centers at Ahmedabad, Bangalore, Cuttack, Gwalior, Allahabad, Chennai, Trivendrum, Hyderabad, Patna and Nagpur. In-house research is conducted at several premier institutions, such as, the All India Institute of Medical Sciences (New Delhi), the Post Graduate Institute of Medical Education and Research (Chandigarh) and Jawaharlal Institute of Post Graduate Medical and Research (Pondicherry). Table 5.3 provides information on grant-in-aidmeant primarily for extramural research during the past five years. The grant-in-aid for extramural research, which was around Rs.11 crore in 1998-99 went upto Rs.27 crore in 2003-04. Basic clinical and operational

Table 5.3 : Expenditure on Sponsored Research –
Ministry of Health & Family Welfare (Rs/Lakh)

S.No. Scheme	*1998-1999 (A)*	*1999-2000 (RE)*	*2000-2001 (RE)*	*2001-2002 (BE)*	*2002-2003 (BE)*	*2003-2004 (BE)*
I. Grant-in-Aid for Research						
1. ICMR	1109.36	1100	1500	1800	2000	2700
2. IIT, Delhi	—	20	50	50	50	45
3. Gandhi Gram Inst.	—	81.80	175	200	114	114
4. Indian Medical Ass.	—	12	15	20	25	25
5. Population Res. Centre	347.50	785	690	—	—	—
II. Grant-in-Aid for Library						
1. National Medical Lib.	483.63	540	529.12	670.59	750	755

BE = Budget Expenditure; RE = Revised Expenditure, A=Actual.

Source: Detailed Demand for Grants, Ministry of Human Resources and Development, Government of India (Several Issues).

research studies relevant to major health problems have been the focus of research programmes by the ICMR. The private sector has been investing mainly in research in the pharmaceutical sector.

According to the Tenth Plan document there is a pressing need for health system research for improving the service delivery and aimed at improving access to technological advances. The Indian Medical Association has been also doing service to the society through raising public awareness vis-à-vis medical ethics through publication and through organizing seminars and conferences. The different types of fellowships offered by the ICMR and the eligible conditions etc. may be seen at Appendix I.

Ministry of Agriculture

The Indian Council of Agricultural Research (ICAR), under the *Department of Agriculture and Research (DARE)* in the Ministry of Agriculture, is the apex body *for coordinating research* in the field of agriculture, soil and water conservation, animal husbandry, fisheries, dairying and forestry. While it supports the various specialized Central Institutes and State Agricultural Universities for in-house research on a regular basis, it also supports independent extramural research through grant of research assistance and fellowships. Grant-in-aid offered for research by the ICAR in different areas of crop husbandry, horticulture, animal husbandry, fisheries, soil conservation and forestry during the past five years may be seen in Table 5.4.

Grant-in-aid has been highest for 'crop husbandry' and included researches being conducted on pulses, oilseeds and maize. Funds allocated for crop husbandry includes receipts from the cess levied on agricultural produce. Two separate enactments exist in this direction, namely, the Agricultural Produce Cess Act, 1940 and the Agricultural Produce Cess, 1966. The objectives of imposing these cess have been primarily to raise resources for supporting research studies on specified agricultural commodities. The total amount incurred on research on 'crop husbandry' during the period 1998-99 to 2003-04 ranged from around Rs.348 crores to Rs.541 crores per annum. This amount also included the 'external aid' component received for conducting research.

Most of these researches are conducted through experts belonging either to the State Agricultural Universities or to the specialized

Table 5.4 : Expenditure on Sponsored Research – *Ministry of Agriculture (Rs/Lakh)*

S.No.	*Scheme*	*1998-1999 (A)*	*1999-2000 (A)*	*2000-2001 (A)*	*2001-2002 (A)*	*2002-2003 (RE)*	*2003-2004 (BE)*
1.	ICAR (Crop Husbandry)	34854*	41920*	40511*	56492.24*	504800	54100*
2.	ICAR (Animal Husbandry)	3700	4220	4464	6580.75	6150	6700
3.	ICAR (Forestry)	775	1000	961	1168.62	—	—
4.	ICAR (Soil & Water Conserv.)	220	230	290	350.58	5950	6100
5.	ICAR (Fisheries)	2100	2400	2534	3064.94	2620	2800
6.	ICAR (Dairy Development)	500	580	690	688.87	—	—

BE = Budget Expenditure; RE = Revised Expenditure. A=Actual

* : This is inclusive of External Aid.

Source : Detailed Demand for Grants, Ministry of Agriculture and Co-operation, Government of India (Several Issues).

institutes in the field of agriculture. The different types of research assistance/fellowships offered by ICAR, their duration and the eligibility criteria etc. may be seen at Appendix I. Besides, the research programmes coordinated by ICAR, in-house research in agriculture is carried out by several institutions such as the Soil and Water Conservation Research Institute (SWCRI) and the All India Soil and Land use Survey Organization (AISLO) etc. The Commission on Agricultural Costs and Prices (CACP) is another apex institution coordinating estimation of cost of production data of principal crops on a regular basis. The main task of CACP is to make recommendations to the Government of India in regard to the Minimum Support Price (MSP) for different crops. These recommendations in turn are based on the data relating to average cost of production of the select crop in the different regions, the input-output ratios, the inter-crop price parity and the cost of living etc. Different specialized institutions and universities are associated in these estimations.

Table 5.5 : Expenditure on Sponsored Research – *Ministry of Environment and Forests (Rs/Lakh)*

S.No. Scheme	*1998-1999 (BE)*	*1999-2000 (BE)*	*2000-2001 (BE)*	*2001-2002 (BE)*	*2002-2003 (BE)*	*2003-2004 (BE)*
I. Grant-in-Aid for Research						
1. Centers of Excellence	515	600	600	650	800	800
2. Professional Societies	10	10	10	10	10	—
3. R&D	275	275	240	240	240	310
4. Integrated Action Oriented Research and Extension	40	40	45	45	45	60
II. Seminars and Training						
1. ICFR	200	210	210	215	250	250
2. Seminars and Symposia	25	25	25	25	—	—

BE = Budget Expenditure; RE = Revised Expenditure; A=Actual

Source : Detailed Demand for Grants, Ministry of Environment and Forests, Government of India (Several Issues).

Ministry of Environment and Forests

The Ministry of Environment and Forests, Government of India has been encouraging both in-house and extramural research in areas such as *eco-development, restoration of damaged eco-systems and eco-regeneration of degraded areas*. The Indian Council of Forestry Research and Education (ICFRE), the Forestry Survey of India (Dehradun), the Institute of Wood Science and Technology (Bangalore), the Institute of Decidious Forests (Jabalpur), the Institute of Rain and Moist Decidious Forests Reearch (Jorahat), the Institute of Forest Genetics and Tree Breeding (Coimbatore), the Forest Research Institute (Dehradun), and the G.B. Pant Himalayan Institute of Environment and Development (Kullu and Gangtok) are the apex bodies under the Ministry to co-ordinate such research programmes

Extramural research is carried on primarily by the ICFRE and the research institutes under the ecological generation programme of the Ministry. Table 5.5 provides information on grant-in-aid extended to 'centres of excellence' and 'professional societies' during the period

1998-99 to 2003-04. Grants-in-aid for extramural research amounted to Rs.10 crores during 2003-04. The ICAR under the Ministry of Agriculture and Cooperation has also been incurring expenditure on forestry research in the range of Rs. 7 to 11 crores per year; the funding, however, appears to have ceased since 2002-03.

Funds have been also provided for organizing seminars and symposia during the different years. The different types of fellowships offered by the ICFR, the duration of fellowship and the eligibility criteria etc. may be seen at Appendix I.

Ministry of Rural Development

The Ministry of Rural Development has also been emphasizing on research (a) for enhancing the effectiveness of its various poverty alleviation programmes (b) for encouraging greater use of technology in rural areas for higher growth and (c) for conducting studies on land reforms as well as for scientific management of transaction in land. Table 5.6 provides information in regard to grant-in-aid provided for agrarian studies, research studies on land reforms, monitoring and evaluation of various schemes under implementation through DRDA (District Rural Development Agency) and for dissemination of latest technology for rural development. The National Institute of Rural Development (NIRD), Hyderabad is the preferred institute for carrying out research studies for the Ministry of Rural Development. The research study pertaining to land reforms was, however, conducted by the Lal Bahadur National Academy of Administration (LBSNAA), Mussourie (Table 5.6).

The Council for Advancement of People's Action and Rural Technology (CAPART) is the apex body under the Department of Rural Development (Government of India) for providing grant-in-aid to *voluntary organizations* for implementing rural development projects and for dissemination of latest technology in rural areas. The nature of financial assistance provided by CAPART and the eligibility criteria for voluntary organizations for availing such assistance may be seen at Appendix I.

Planning Commission

The Socio-Economic Research (SER) Unit is the nodal division in the Planning Commission for supporting studies relating to planning

Table 5.6 : Expenditure on Sponsored Research –
Ministry of Rural Development (Rs/Lakh)

S.No. Scheme	1998-1999 (A)	1999-2000 (A)	2000-2001 (A)	2001-2002 (RE)	2002-2003 (RE)	2003-2004 (BE)
I. Grant-in-Aid for Research						
1. Agrarian Studies	36	Nil	48	38.67	48	48
2. Land Reforms	25	Nil	Nil	Nil	Nil	Nil
3. Monitoring & Evaluation of DRDA schemes	61.84	26.83	55:64	4.90	50	50
4. Rural Housing		26.83	55.64	4.90	50	50
5. Social Security and Welfare	2.31	5	5	5	—	—
6. Self-employment project			56.99	49.81	50	40
7. Swaranjayanti Gramin Rozgar Yojana			12.96	23.36	20	30
8. Village Panchayat			15.55	27.25	15	15
9. CAPART*	1200	1300	1300	3000	3000	5000
II. Seminars and Training						
1. Conferences & Seminars	31	116	130	24.73	30	Nil
2. Training of Rural Youth for self-employment	4500	5495	Nil	Nil	Nil	Nil
3. Panchayat Development and Research		100	100	96.02	300	120

BE = Budget Expenditure; RE = Revised Expenditure; A=Actual

DRDA=District Rural Development Agency.

* This includes outlays incurred on carrying out rural development projects through VO's.

Source : Detailed Demand for Grants, Ministry of Rural Development, Government of India (Several Issues).

and development. The Commission has been conducting (extramural) research in areas that need immediate attention through reputed research institutes. Grants-in-aid is also provided for in-house research to the Institute of Applied Manpower Research (IAMR) on a regular basis for conducting studies which are of interest to Planning Commission. The Programme Evaluation Organization (PEO) is another wing in the Planning Commission which conducts ex-post evaluation of welfare programmes being implemented generally at the national level. The PEO has brought out evaluation studies for a large number of government programmes and projects. The PEO generally carries such studies on the request made by the Ministry/ Parliament. The recommendations made by the PEO are given due cognizance by the Government.

Apart from such exigencies, the Commission supports requests for financial assistance made by research institutes and research scholars. The relevant subject divisions of the Planning Commission initially examine the applications for financial support. Based on their comments the Research Committee under the chairmanship of the Secretary, Planning Commission grants the financial assistance. While the ceiling for grant-in-aid for research studies conducted by institutes/universities is Rs.8 lakh (in exceptional cases Rs.15 lakhs), it is Rs.2 lakh for individual research scholars. Table 5.7 provides information on the grants -in -aid offered for extramural research during the last five years. Expenditure incurred/sanctioned on grant-in-aid for research during 2002-03 and 2003-04 amounted to Rs.1.5 crore per annum. The Planning Commission has been also entertaining

Table 5.7 : Expenditure on Sponsored Research -
Ministry of Planning/Planning Commission

(Rs.in '000)

S.No. Scheme	*1998-1999 (A)*	*1999-2000 (A)*	*2000-2001 (A)*	*2001-2002 (RE)*	*2002-2003 (RE)*	*2003-2004 (BE)*
Grant-in-Aid for Research						
1. Grant-in-aid*	1,33,76	69,52	9,965	1,24,79	1,50,00	1,50,00

BE = Budget Expenditure; RE = Revised Expenditure; A=Actual.

*Grant-in-aid to Universities and Research Institutions

Source : Detailed Demand for Grants, Ministry of Planning, Government of India (Several Issues).

requests for organizing workshops and seminars. The eligibility criteria for institutions and scholars for availing financial support for different objectives from the Planning Commission may be seen at Appendix I.

Summing Up

The data above capture the flow of funds for sponsored research, from the Government of India to the various autonomous research councils and research institutes, during the past five years. The University Grants Commission (UGC) and the All India Council of Technical Education (AICTE), under the Ministry of Human Resources Development and the Indian Council of Agricultural Research (ICAR), under the Ministry of Agriculture (Government of India) are the major 'funding windows' disbursing grant-in-aid for extramural research in the range of hundred to five hundred crores every year. Annual expenditure on extramural/sponsored research funded by the Council of Scientific Research (CSIR), the Indian Council of Medical Research (ICMR), the Indian Council of Social Science Research (ICSSR) and the Indian Council of Forestry Research (ICFR) etc., on the other hand, have so far been in the range of fifty to seventy crores per year Similar flow of funds for sponsored research by the private sector is not easily available. In view of the heightened competition in the market arising from globalization, the industrial sector needs to be forward looking and ought to collaborate with universities and other academic institutions for improving their products.. The Planning Commission has been providing support for extramural research for socio-economic studies since the beginning of planning in India; the Commission could significantly increase the plan allocations for research.

Annexure 5.1

Field Perspective – A Case Study

The case study is based on a sample survey conducted for the Patna University. The survey includes both the disciplines of 'physical sciences' and 'humanities'. The questionnaire that was designed to elicit information in regard to the difficulty/inability in accessing the different funding windows, may be seen at Appendix III. The level of research activities, viewed from the angles of both demand and supply in the University is noted to be low. It also emerges that even when Ph.D was made a condition for promotion of lecturers, the motivation for research was found lacking. It has, therefore, been suggested that the approach of teaching for examinations should be replaced by teaching for knowledge. Moreover, the following kinds of difficulties are commonly experienced in carrying out research:

1. Lack of books, journals and modern facilities in libraries tailored to suit the specific needs of researchers,
2. inadequacy of laboratories having modern facilities,
3. lack of educational programmes relating to research.

Grant-in-aid for research is at best playing a passive role in promoting research. Information regarding diverse 'funding windows' other than those of the Government is not reaching the people keen on pursuing research. *Administrative irritants also need to be addressed urgently*. Whenever any fund is sanctioned by the funding agency, such as, the U.G.C., Department of Science & Technology or the Ministry of Environment and Forest (Government of India), the sanctioned amount is first sent to the Registrar of the University and from the Registrar to the Supervisor, which normally takes three to four months. The next round of sanction is, furthermore, made based on the provision of 'utilization certificate' sent from one level to another level of official hierarchy; and all this red tapism involves delays and wastage of time. This highly centralized system could be improved through transferring funds directly to the Supervisor, the Head of the Department or to the Principal of the College.

Following recommendations could be further considered for improving the situation, namely

1. Well-conceived seminars/workshops should be organized (a) for providing relevant information to those who are keen to do research

and are in need of financial assistance, (b) for identifying specific problems faced both at the micro and macro level and (c) for preparing action plans for addressing the problems.

2. Action Plans might be regularly reviewed and this should be a continuous process. As the reviews would provide new insights into the problems, subsequent Action Plans ought to be accordingly formulated.
3. A system of evaluating research output may be put in place for recognizing good works in the form of 'prize essays'/reward publications.

It was also felt that focus on research is mainly concentrated in Delhi and a few select places. This should be replaced by encouragement to research across the entire system. This view is very similar to those of the National Policy on Education (1986).

Chapter 6

Future of Research in India

Both in-house and extramural research showed an up trend during the 1990's and that trend has continued in the new millennium. There are the usual demand and supply side to research. On the supply ride there is a pool of talented young men and women in the country. Competition in the global market, the need to reduce costs and the desire to introduce new products or differentiate products are compelling factors on the demand side. These factors, in turn, are pointers to India emerging as an important centre of research activity. The demand for manpower in the area of science and technology alone is estimated to equal a workforce of 50,000 during the Tenth Five Year Plan (2002-07).

At the same time, there is need to bring about a change in mindsets. Young pioneers need to be assured of a career in research and should be encouraged to think in terms of setting up their own establishments with the help of 'venture capital'. Markets are very often psychology and culture driven and they provide good scope for socio-economic research. At the level of Government, there is need to further strengthen the research infrastructure in the country in the form of good libraries, network of laboratories and institutions of higher learning. Use of computers has revolutionized compilation, tabulation and analysis of data and its utility needs to be sufficiently recognized.

6.1 Prospects of Scientific Research

Future prospects of scientific research in India may be based on the past trends vis-a-vis (a) the overall investment in R&D, (b) investment on R&D in the government sector, (c) investment on R&D in the private sector, (d) investment in R&D by global firms in India and (e) growth in the number of patentable innovations. An attempt has been made in the paragraphs below to discuss each of these trends.

Table 6.1 : Expenditure on R&D in India

Items	*1970-71*	*1980-81*	*1990-91*	*2000-01*
GNP at Current Prices (Rs. million)	419380	1305250	5034090	19003100
Expenditure on R&D at Current Prices (Rs. million)	1396.4	7605.2	39741.7	178117.0
Expenditure on R&D as per cent of GNP	0.33	0.58	0.79	0.93

Source : Hand book of Industrial Statistics, Ministry of Commerce & Industry, Government of India, New Delhi, 2002.

6.1.1 Investment in R&D

There is a noticeable increase in R & D expenditure during the past four decades, both in absolute terms and as percentage of GNP (Table 6.1). Expenditure on R&D thus increased several times from about Rs.40 crores in 1970-71 to Rs.17,812 crores in 2000-01. Even, if necessary correction is made for inflation, expenditure on R&D has gone up significantly over these years.

An analysis of R&D expenditure by different sectors, moreover, shows that private sector investment saw the highest growth during this period (Table 6.2). The share of private sector in total R&D investment, which was 10 per cent during 1970-71 went up to 23 per cent in 2000-01.

Table 6.2 : Expenditure on R&D by Sectors

(Rs.in million)

Sectors	*1970-71*	*1980-81*	*1990-91*	*2000-01*
Central Sector	1124.7 (80.5)	5804.9 (76.33)	30582.7 (76.95)	118357.6 (66.45)
State sector	125.8 (9.01)	593.4 (7.80)	3659.2 (9.21)	13505.6 (7.58)
Private Sector	145.9 (10.45)	1206.9 (15.87)	5499.8 (13.830)	40703.6 (22.85)
Total	***1396.4***	***7605.2***	***39741.7***	***178117.0***

Note : Figures in parenthesis represent percentage of the total.
Source : Ibid.

Private sector has emerged as the major (R&D) investor in areas such as biotechnology, drugs & pharmaceuticals, electrical and electronics equipments, chemicals, transportation, mechanical, engineering and 'information & technology' (Table 6.3). The share of private sector investment, ranged from 75 per cent to 99 percent, of total investment in these emerging areas. Public sector, nevertheless continues to be the major (R&D) investor in 'defence industries'. Share of public sector in defence industries may also come down in future as the Government has opened (dereserved) this sector for private sector investment.

6.1.2 R&D in Government sector

Although the share of Government expenditure on R&D, out of the total expenditure incurred, has gone down, there has been marked

Table 6.3 : R&D Investment by Emerging Areas, 1998-99

(Rs. million)

S.No.	*Industry Group*	*Public Sector*	*Private Sector*	*Total*
1.	Biotechnology	6.96 (0.13)	51956.3 (99.87)	52025.9
2.	Drugs & Pharmaceuticals	49.01 (1.29)	3725.95 (98.70)	3774.97
3.	Defence Industries	2188.17 (99.94)	1.42 (0.06)	2189.59
4.	Electrical & Electronic Equipments	309.09 (15.45)	1691.27 (84.55)	2000.30
5.	Chemicals	270.52 (13.86)	1680.85 (86.14)	1951.38
6.	Funds	1558.80 (99.49)	8.01 (0.51)	1566.81
7.	Transportation	73.03 (4.78)	14552.99 (95.22)	15283.29
8.	Telecommunication	611.39 (71.14)	248.08 (28.86)	859.48
9.	Misc. Mechanical Engineering	52.22 (12.94)	351.26 (87.06)	403.49
10.	Information & Technology	91.38 (23.83)	292.13 (76.17)	383.51

Note : Figures in parenthesis are as percentage of the total investment in the specific sector.

increase in R&D investment by the Government in absolute terms (Table 6.4). In areas such as space, atomic energy, defense and ocean development, expenditure on R&D has taken the 'mission mode'. R&D expenditure, in most of these areas under the 'mission mode', increased more than 200% during the period 1990-91 to 1998-99.

Due to the thrust given to development of defence research and the need to develop the requisite 'technology infrastructure', India has emerged as one of the few countries in the world capable of making highly sophisticated weapon systems, including missiles of various descriptions and multi-barrel rocket systems. The country has also developed low-level tracking radar, night vision devices, and sophisticated ship sonar systems. An indigenous light combat vehicle and a remotely piloted vehicle was also successfully tested. Significantly, the spin offs from defense research to commercial R&D and product development is also increasing.

Table 6.4 : R&D Expenditure by Major Government Scientific Agencies.

(Billion Rupees)

Agency	*1990-91*	*1998-99*	*% increase*
Department of Space	3.9	15.2	292.4
DRDO	6.8	23.0	238.2
Indian Council of Agricultural Research	2.8	8.4	205.5
Department of Ocean Development	0.3	0.8	205.0
Department of Atomic Energy	2.8	8.4	203.7
CSIR	2.5	7.1	184.0
Department of Science & Technology	1.2	3.0	149.5
Ministry of Environment & Forests	1.6	3.8	133.2
Department of Biotechnology	0.4	0.9	128.4
Indian Council of Medical Research	0.4	0.9	93.7
Ministry of Information Technology	0.3	0.6	88.0
Ministry of Non-Conventional Energy Sources	0.2	0.1	-44.1
Total	***23.1***	***72.2***	***212.3***

Source : Based on Department of Science & Technology (2002)

Indian space programme which dates back to 1963, formally took off in 1972 as 'a national space programme' with well defined objectives of applying space technology to communications, meteorology and resources management. India thus has a credible capability in space science and technology, including design and construction of satellites and satellite launch vehicles. The first successful flight of GSLV-D1, development of PSLV with the successful flight of PSLV-C1, the launches of IRS-1D, P4 (ocean sat) and TES, INSAT-2E and INSAT-3B are some of the significant achievements under the space programme. The INSAT range of satellites has revolutionized, communication, education, disaster warning system and watershed management systems.

The nuclear power programme under the Government has reached an advanced stage of implementation and R&D in this area has resulted in India mastering all aspects of this difficult technology. The entire range of technologies from processing of raw materials to design and construction of large nuclear reactors have all been successfully developed and tested. There have been considerable achievements made in the field of computers and super computers. The path-breaking developments in parallel computing have broken the supremacy of developed countries. India has unveiled the new, PARAM 10000 supercomputer, with a capability of 100 Gigaflops (100,000,000,000 mathematical operations per second). The open frame architecture of C-DAC's PARAM 10000 places India in the select league of nations, which are expanding the frontiers of supercomputing to teraflop range.

The Council of Scientific & Industrial Research (CSIR) is the premier national R&D organization, having a network of 40 laboratories and 80 field centres for scientific and industrial R&D. The organization is being revamped with the changing national and international scenario. The white paper titled "CSIR 2001, Vision and Strategy" prepared in 1996 set the motion for the restructuring of CSIR. The document recommended for re-engineering the organizational structure with a view to infuse customer orientation and market responsiveness, *linking R&D to market place through alliances and networking,* stimulating intellectual property management, investing in select high quality science and refurbishing the ageing human capital. The new approach has paid dividends as it led to a total external cash-flow of around Rs.1000 crore during the period 1997-2001. The industrial production catalyzed is stated to be of the order of over Rs.17, 000 crore, for the same period.

In order to consolidate the achievements further, the Government has identified following areas under the mission mode programmes during the Tenth Plan period. They are mentioned below department wise:

Department of Atomic Energy

1. Advanced Heavy Water Reactor (AHWR) and development to technology for utilization of Thorium for third stage of nuclear power programme.
2. Technology demonstration and transfer of water de-salination.
3. Nuclear medicine.
4. Application of irradiation technology of farm products.

Department of Space

1. Operationalization of National Natural Resource Management System (NNRMS) for application of remote sensing technology in developmental programme.
2. Technology development for future generation of launch vehicle/ development of advanced launch vehicle technology.
3. Development of all weather remote sensing technology.

Department of Science & Technology

1. Technology of bamboo products.
2. Drugs & Pharmaceutical research.
3. Seismology
4. Nano Science & Technology

Department of Biotechnology

1. Genomics covering human plants animals and microbes with emphasis on drug targeting.
2. Development of new drugs and molecules from important medicinal plants.
3. Development of New Generation vaccines.

Department of Ocean Development

1. Exploration and technology development of gas hydrates.
2. Large scale technology demonstration of ocean thermal and energy convergence plant.

Council of Scientific and Industrial Research

1. Development of technology for control of Asthma caused due to pollution.
2. Leather technology
3. Design & Development of civil air craft-stretched SARAS and HANSA.
4. Study of Mesozoic sediments for hydrocarbon exploration in coordination with ocean development.
5. Development of microwave tube technology for ultra high frequency communication for large scale application.

Apart from these, CSIR has identified 24 Mission Mode Programmes in the areas of specialized Aerospace Materials; Photonics and Opto-electronics; New Generation Fuels and Lubricants; Custom Tailored Special materials etc. and Core Programmes in the areas of developing and sustaining high S&T for National Aerospace Programme; Electronics for Societal purposes; Advanced facility for Safety Evaluation of Genetically modified/Engineered Drugs; advanced manufacturing technology etc.

6.1.3 R&D in the Private sector

During the pre-liberalization period, investment in R&D by the private sector was very low. This is attributed to the low level of competition as firms enjoyed a fairly high degree of protection on account of high customs duties. While the share of private sector in the total R&D expenditure in 1990-91 was around 14 per cent (Rs.550 crore), it went up to 23 per cent (Rs.4070 crore) in 2000-01. The large high tech firms and firms with international outlook have brought about the real change as is evident from the performance of the top 20 domestic firms in respect to R&D investment (Table 6.5). The average R&D intensity (ratio of investment in R&D to annual sales) works out to more than 2 per cent for these firms. The firms in pharmaceuticals, automobile and software sectors especially, have been the 'front runners'.

Table 6.5 : R&D Expenditure of Top 20 Firms, in 2002

(Rs. Million)

Firm	*Revenue Expenditure*	*Capital Expenditure*	*Sales*	*R&D Intensity (%)*
Dr. Reddy's Laboratories Ltd.	980	0	17118	5.7
Tata Engineering	923	0	87101	1.1
Bharat Heavy Electricals Ltd.	796	75	76511	1.1
Bharat Electronics Ltd.	783	112	19473	4.6
Reliance Industries Ltd.	749	152	57120	1.6
Oil & National Gas Corpn. Ltd.	709	93	23755	3.4
Mahindra & Mahindra Ltd.	689	0	39345	1.8
Ranbaxy Laboratories Ltd.	647	124	23625	3.3
I T I Ltd.	497	83	24322	2.4
Steel Authority Of India Ltd.	481	17	16644	3.0
Indian Oil Corpn. Ltd.	450	235	117107	0.6
Cipla Ltd.	367	100	14008	3.3
Wockhardt Ltd.	360	41	64942	0.6
Bajaj Auto Ltd.	337	36	41721	0.9
Hughes Software System Ltd.	334	0	23488	1.4
Ramco Systems Ltd.	268	31	9384	3.2
B P L Ltd.	264	0	12979	2.0
Maruti Ugyog Ltd.	261	195	90867	0.5
Lupin Ltd.	250	285	96354	0.6
T V S Motor Co. Ltd.	245	53	19305	1.5
Mean				***2.1***

Source : Based on Bowender (2003)

The pharmaceutical industry, which was practically non-existent at the time of independence has emerged as one of the most competitive producers of therapeutics in the world. It has contributed greatly to improving the standards of healthcare in the country and making modern medicine available to the people at affordable prices. India is a net exporter of pharmaceuticals today and meets more than three quarters of its requirements of bulk drugs and almost the entire requirement of formulations. Agro-chemicals is yet another area of success for Indian science and technology (S&T). The industrial sector, which was predominantly dependent on imports until the early seventies, is self-sufficient today in all matters of technology and production. In the area of petroleum refining and petrochemicals also, the country has made impressive strides. The Indian leather industry has been transformed from a mere exporter of raw hides a few decades ago to a vibrant, modern industry that ranks among the top five export earners. Research and development (R&D) in this sector has contributed to the entire range of technical activities from environmentally clean leather processing technologies for curing, tanning and finishing to process automation and modernization of tannery operations, and novel product design.

6.1.4 Growth of R&D Centres by Global Firms in India

Recognizing a reasonably well developed R&D infrastructure, education and training facilities and cost effective English speaking intellectual resource base, many global firms have either established or are in the process of setting up R&D centres in India. As many as 77 global firms are noted to have set up R&D centres in India during the last three years (Table 6.6). These firms include multinationals

Table 6.6 : R&D Centres of Global Firms Operating in India

Location	*Firms*
Bangalore	Akzo Nobel, Bell Labs, Delphi, DuPont, Seagate, Daimler Chrysler, GE Industrial Systems, GE Medical Systems, General Electric, General Motors, Hekel Loctite, Hewlett, Packard, Honeywell, Texas Instruments, Astra Zeneca, Unilever, CISCO, Nokia, Philips, Motorola, Tyco Electronics, Siemens, Lucent, Sanyo, SAP, Gulf Oil, Exxon, Snecma, 3Com, Pepsico, Nortel, Sharp, Unilver, Cadence.
Delhi	AVL, IBM, DuPont, Eli Lilly, SGS Thompson, Adobe, LG Electronics, Honda, Daewoo, Ikos, Nestle, Cadence
Hyderabad	Motorola, Oracle, Vaanenburg, Microsoft, Intergraph, GE Industrial Systems, Arsin, Ericsson, Bell Labs
PuneMumbai	Whirlpool, GE Plastics, Emerson Electric, Cummins, Syntel, Colgate Palmolive, Hoechst, McDonalds, Pfizer, I-Flex, Shimadzu, Scala, Unilever
Goa	Seagram
Chennai	Ford, Hyundai, Cognizant, DuPont

Source : Ibid.

such as the General Motors (GM), the General Electronics (GE), Motorola, Texas Instruments, International Business Machine (IBM), HP, Dell, Nortel Adobe, SGS Thompson, Mentor Graphics, Monsanto, Daimler-Chrysler, Siemens, Exxon, etc. The entry of global firms with high R&D capability is bound to enhance the product development capability in the country. It is supposed to make a bigger impact on account of the knowledge network created at industrial hubs in the country such as Bangalore, Hyderabad, Pune, Delhi, Chennai and Mumbai.

6.1.5 Growth in the Number of Patents

One of the most popular measures to assess the status of R&D in any country is the number of patents filed and granted in the US Patents and Trademarks Organisation (USPTO). There has been a marked increase in patentable innovations in the post-liberalization period. The new patent registrations, patents filed by the global firms operating in India and the number of Indian patentees, have grown many folds. There has been a ten-fold increase in the patents filed in the USPTO by Indian firms during the period 1996-2002 compared to the earlier period of 1991-95 (Table 6.7). The firms such as Dr. Reddy's Laboratories, Ranbaxy, Panacea Biotech and a large number of individual patentees, contributed in a big way to this development. There has also been a marked increase in the US patents granted to inventions from CSIR Labs in recent years. The CSIR was granted 61 and 128 patents in 2001 and 2002 respectively. These results show that the focus of research is shifting to patentable innovations with an eye on the market.

It is quite clear from the above that while the pre-liberalization period saw R&D research directed towards assimilation of imported technology and development of indigenous technology, the post liberalization R&D endevour is *to give India a competitive position in the new global economy through increasing India's share in high-tech products. There is nevertheless a need to strengthen the* enabling environment for R&D in terms of improving the technology infrastructure, enhancing the industry-institution linkages, development of science and technology manpower and providing fiscal incentives for encouraging both extramural and in-house R&D.

Table 6.7 : Patent Granted by USPTO to Indian firms/ institutions

Agency	*No. of Patents.*	
	1991-95	1996-02
Total Indian Assignees	54	583
CSIR	25	276
Dr. Reddy's Labs Ltd.	-	28
Ranbaxy Labs Ltd.	5	29
Panacea Biotech	-	13
NRDC	-	8
IPCL	-	7
Indian Oil Corporation Ltd.	-	14
National Institute of Immunology	-	9
Individual Patentees	29	96
Patents taken by Non-Indian R&D labs Operating in India with Indian Inventors		
Mansanto Co.	3	-
Texas Instruments Inc.	11	63
General Electric Co.	1	45
IBM Corp.	-	31
Lucent Technologies Inc.	-	13
Hoechst Aktiengesselschaft	30	10
Intel Corp.	-	7
Patents: An International Comparison		
Brazil	188	359
India	54	583
China	148	735
Israel	1032	3555
Korea	3690	20744
Taiwan	2240	18031

Source :Ibid.

According to the National Science Foundation (NSF), USA an independent agency specializing in assessing the quality of technological infrastructure in different countries, India improved its score from 33 in 1993 to 37 in 2003, on a 0-100 *point scale.* Table 6.8,

Table 6.8 : Technologival Infrastructure Indicator

Sl No..	*Country*	*1993*	*2003*
1.	India	33.0	37.0
2.	Malaysia	34.2	28.8
3.	Thailand	26.7	23.9
4.	Argentina	25.5	31.0
5.	Brazil	41.6	35.4
6.	Mexico	25.2	28.7
7.	Israel	NA	52.3
8.	China	38.6	55.2
9.	Japan	92.7	NA
10.	Korea	42.6	NA

Source : National Science Foundation (1995) and (2004)

moreover, suggests that India ranks well above a number of developing countries in terms of technological infrastructure. Its achievement in technological infrastructure is, however, low compared to the industrially advanced countries like Germany, Japan and also China.

The following method of quantifying the technological infrastructure has been used:

Quantifying the Technological Infrastructure

Five variables are used for developing this indicator, which evaluates the institutions and resources that contribute to a nation's capacity to develop, produce, and market new technology. The important variables considered are:

- Number of scientists engaged in R&D.
- National purchase of electronic data processing equipment.
- Nation's capability to train citizens locally in academic science and engineering, the ability to make effective use of technical knowledge and the linkages of R&D to industry.

Technology infrastructure (T1) = (a+b/2) +c+d+e+f

Where:

(a+b) = output of indigenous science and engineering;

c = survey data assessing linkages of R&D of industry

d = ability to make effective use of technological knowledge

e = national purchases of electronic data processing equipment

f = the number of scientists and engineers involved in research

(Note: Raw data are transformed into scale of 0-100 for each indicator component and then averaged to generate comparable indicators with a 0-100 range. For survey items, 100 represents the highest response for each category for each question; for statistical data, 100 typically represents the value attained by the country with largest value among all the countries (28) included in the larger study. Each item carries equal weight.)

From the point of view of alluring global investments into R&D, therefore, there is any urgency to pay greater attention to improving the technological infrastructure in the country. Paragraphs below provide a broad outline of *the institutional infrastructure* existing in India for the development of R&D.

Technological Infrastructure

The institutional infrastructure for technological development in India can be divided into six main groups namely:

1. Institutions for policy planning and administration, which are mainly the following:
   a. Department of Science & Technology (DST)
   b. Department of Scientific & Industrial Research (DSIR)
   c. Department of Biotechnology
2. Institutions related to generation of Technology, such as,
   a. CSIR laboratories
   b. Scientific and Industrial Research Organisations (SIDO)
   c. Laboratories under various ministries and specialized labs such as the Centre for Development of Advanced Computing (C-DAC) and Centre for Development of Telematics (C-DOT)
3. Institutions transferring the generated technologies to the industry, for example,
   a. NRDC
4. Financial institutions for Technology Development
   a. Development Finance Institutions such as IDBI and ICICI
   b. Technology Development Board
   c. Government/ private sector venture capital institutions
   d. TIFAC
5. Institution for information on technology and forecasting of technology, namely;

   a. TIFAC
   b. NISSAT
6. Standards setting and quality certification institutions, for example,
   a. Bureau of Indian Standards (BIS)

6.2. Prospects of Socio-Economic Research

Research is useful both in the realm of physical sciences and in the socio-economic field. In fact, the aim of research in physical sciences is to ultimately achieve social good/ social objective. Sufficiency-insufficiency of resources may, however, make people live with alternative technologies. The level of literacy and illiteracy has also been observed to make a great deal of difference in adopting known technologies. Inequitable distribution of social wealth and poor service delivery, similarly, prevent everyone to benefit equally from various inventions made in the field of physical sciences. The needs and preferences of a child, a man and a woman, moreover, vary as the needs of the young and the old are also at variance. A proper assessment can accordingly be made through conducting surveys in the desired direction. Governments of welfare states spend huge sums of money on different programmes relating to health, education and infrastructure. It is possible to bring down the costs and increase efficiency through evaluation of these programmes and introducing efficient management systems.

Activities relating to production, distribution, exchange and distribution in an economy likewise call for regular monitoring for necessary policy initiatives by the government. India has had fairly a good database in regard to population statistics, livelihood status, landholdings pattern, agricultural and industrial production, public finance data, data relating to money supply and prices, trade and commerce etc. The country has a large network of institutions, such as, the Registrar General of India (RGI), the Central Statistical Organization (CSO), the National Sample Survey Organization (NSSO), the Directorate of Economics and Statistics (DES), the Department of Industrial Production (DIP), the Directorate General of Commercial Intelligence and Statistics (DGCIS), the Reserve Bank of India (RBI) and the Registrar of Companies etc. These are national institutions and they perform their task in association with the federal governments. The field offices of each of these institutions generally exist in each state.

The Office of the Registrar General of India, for instance, collects information on the different demographic features, such as, headcount of the population, age profile, male-female ratio, life expectancy and occupational pattern etc. The CSO estimates the national income, consumption expenditure, savings-investment, changes in stock and per capita income etc in association with the Directorate of Economics & Statistics of each state. The NSSO, on the other hand, collects information based on sample surveys on subject matters, which do not have the system of data collected on an annual basis. Besides these, there are host of other institutions that collect socio-economic data as per the requirements on varied social indicators ranging from health to housing. Institutions, such as, the NCAER and NABARD etc in the public sector and ORG, the CMIE, the FICCI etc. in the private sector have been conducting large-scale national surveys in specific areas.

The coverage, reliability and periodicity of data collection for many sectors in India, however, fall short of the standards set in advanced countries. Data on employment statistics for instance are very weak and there is a clear need to bring about necessary improvement in data collection in all such areas. Research scholars belonging to universities and specialized institutes, nevertheless, regularly have been making significant contributions through articles and books on various socio-economic aspects. Dedicated institutions such as the ICSSR, the ICHR, the ICPR and the IIAS and numerous specialized institutions, apart from the universities, are engaged in socio-economic research. The media, both print and visual, has emerged as yet another section of the society very actively engaged in collecting and disseminating socio-economic data. With the reality of several centuries co-existing in India and the visible diversity of the sub-continent, India qualifies very well as an ideal destination for socio-economic research.

Operational Research

The individual firm or a corporation benefits through 'market research' as it enables it to make proper assessment of demand and supply of a 'good' or service. 'Management' has emerged as a useful subject under 'social sciences', and promises to provide solutions to numerous problems faced in the market. It has to-day branched off into different areas such as financial management, personnel/corporate management and sales/marketing management. It has helped produce business leaders through enabling them to be ahead of others. Its success, in the final analysis, depends upon the research techniques it employs. 'Operational Research' is one of the tools widely used by managers. It provides executives with a quantitative basis for taking decisions regarding operations under their control. In regard to a typical firm, 'operational research' helps a firm draw up its plan for the long-range requirements of capital, optimum resource allocation and inventory management as well to account for the investment risks involved. Operational research can be used to test innovative models in the social sector, which are more effective.

Computers as a device have created new potential of recording and collating data and have made it possible to do all kinds of analysis. Information technology has already revolutionized the way people have hitherto lived and the World Wide Web is considered the biggest library of the world. A great deal of benefit also accrues through simple transfer of technology. The multilateral organizations of the UN such as UNICEF, UNDP, FAO, UNESCO, WFP have been playing significant roles in transferring technologies and know-how acquired in the developed countries to the developing countries. In this regard, it may not be far fetched to visualize that soon Indian experts would be working shoulder to shoulder with the tallest from the rest of the world *in larger numbers,* in transferring 'state-of-art-technology' to the developing countries. There is also an ethical aspect to research and this also needs to be actively debated for a secure future.

Chapter 7

Conclusion

The spirit of enquiry is basic to research. Research leads to introduction of new products and services. *Clustering of innovations have been observed to usher long periods of high growth in the (world) economy.* The 'Industrial Revolution' in England, during the 17-18th century and the more recent 'Green Revolution' in India have been the results of scientific research. India spends around one per cent of its GDP on research compared to three percent and more by a number of countries. Demand for R&D, soon after independence, arose for development needs. The demand for R&D to-day, is driven by competition in the market. 'Globalization' is indeed a threat for products and services produced domestically under the (WTO) regime of falling *tariffs*. At the same time, 'globalization' provides opportunities to enterprising firms for going beyond the limits imposed by the national market *through exports*. The *high technology global competition* has critical implications especially for corporate planning and for policy making of national governments.

Whether it is capital goods, consumer durables, electronics goods/ toys or services, the technologically superior products have their own appeal across countries. The reason why some countries have a larger share in world trade than others is that they have more of high technology exports. High technology products, in turn, are defined as those that have an R&D spending equal to or greater than 3.5 per cent of the sales value of products. Data show a positive relationship between R&D investments and export performance for a large number of industries. The most consistent results supporting such relationships have been found in the chemicals, drugs, plastics and machinery industry (Agarwal, 2002). Positive correlation may also be shown to exist between the number of people employed in research activity and exports. For instance, while India's share in world trade is less than one per cent that of China is around four per cent and *the R&D*

personnel engaged in industries in China was observed to be five times that of India during 2000-01. Needless to say, larger sales engineered by (international) trade enable firms to reap economies of scale and thus become more competitive.

The various initiatives taken by the Government of India, since independence, have created a reasonably good R&D infrastructure in the country in the form of a chain of national laboratories and *institutions of advance technical education* etc. The private sector has also created its own R&D units and sees a business opportunity today in research. With over 200 universities, more than 1500 research institutions and around 500 PhD's every year India has the necessary wherewithal to become an important centre for research and development. There is, however, a need to pay *greater attention to technological infrastructure* since India lags behind the R&D infrastructure of the advanced countries. Areas of research in physical sciences encompass agriculture, non-conventional energy sources, atomic energy, electronics super-computer, telecommunication, automobile, ocean, space, biotechnology, health and environment.

What is most important, nevertheless, is *human development* for better economic progress. With two-third people in India dependent on agriculture and residing in rural areas, there is a need to find opportunities of growth in sectors and in regions close to the people. This can be best achieved through socio-economic research. Socio-economic research helps make better assessment of the needs and preferences of people, run a scheme better through 'monitoring' and improve upon an activity through (ex-post) evaluation. Needless to say education, better nutrition, good infrastructure and useful institutions go a long way to bring about a qualitative difference in the lives of people and in the environment around.

Research could, however, be organized in different forms. It could be either 'in-house' or sponsored research'. There is today a general recognition of the need to bridge the gap between 'the stand-alone research institutions' and 'institutions of higher learning'. Sponsored research is indeed a very good way out to address this problem. Whether it is the National Educational Policy (1996) or the Tenth Five Year Plan (2002-07), they all endorse this approach. FDI into R&D has been coming into India by way of alliances (collaboration) with local firms and research institutions. The pace of foreign direct investment (FDI) may significantly go up in future. A new trend is noticeable amongst multinational companies for outsourcing R&D activities through entering into 'contract research' with local firms/ institutions, especially in the area of biotechnology.

Some Indian companies and institutions have also made successful entry into global R&D and have been developing world-class products. The information technology sector has been in the lead in this respect. This sector has also attracted considerable foreign direct investment into India. According to the Tenth Plan document, 'most of the leading international players especially those in very large scale integrated design have set up their design and R&D centres in India' (Planning Commission, TFYP, 2003). The Plan document, nevertheless, observes that the software industry has been mainly concentrating on the service sector and that 'although the Indian professionals and the Indian industry have contributed to the development of intellectual property, the Indian industry owns very few patents' (Planning Commission, TFYP, 2003).

The Central Government and the State Governments have been utilizing services of 'professionals' and senior faculty members of universities/research institutes in their respective areas of specialization for policy formulation. Studies are also conducted through experts and consultants. Research impacts development through bringing the desired synergy between different programmes and through increasing the cost effectiveness of several welfare programmes. Workshops, consultations and seminars are other options to discuss issues and finally make the recommendations. If a comparison were, however, made between in-house research and extramural sponsored research, it would be obvious that extramural research constitutes a small portion of the expenditure incurred on in-house research. It could be argued that flow of funds for research in institutions of higher learning is more a result of adhoc decisions (except the University Grants Commission) and not yet an "act of faith" with the other funding agencies.

Amongst the new initiatives to promote sponsored research in India, perhaps, the efforts towards accrediting universities/departments is the most promising. This simplifies the task of agencies keen to collaborate/fund the project. It is the National Accreditation and Assessment Council set up by the UGC which performs this task for the universities and their colleges on a 'self assessment appraisal' basis, every five years. This could be supplemented/substituted by the practice prevalent in the United Kingdom, whereby it is the select panel constituted (privately) by the different subject specific *academic associations* who do this task of evaluating the quality of research output (and other parameters) on a five point scale each year.

The number of patents registered with domestic or international agencies is another form of certification for the institutions of higher learning. The *quality of research output in the country* also needs to be documented for a correct assessment. Usefulness of research may similarly be greatly enhanced if this is done *with the use and the user in mind.* The high drop out rate (35 per cent) reported for research fellows qualifying in NET, conducted by the University Grants Commission (UGC) is, indeed very worrisome. This only shows that there is not sufficient incentive for research.

The issue of Intellectual Property Right (IPR) has been a much-neglected area in India. Creating a scientific manpower in a traditional society was the prime objective of the R&D policy in India. It failed, however, to make the researchers, research institutes and the corporate world become direct participants in the technology revolution *through securing for them direct benefits for the innovations made.* The IPR helps recover the cost of developing a product (or process) by way of *the patent right* for the 'innovation' and is a device to reward the author or the institution by way of 'royalty' through the device of copyright. There is nonetheless, the negative aspect of IPR as it restricts dissemination of research during the period of patent protection and slows down diffusion of knowledge. The Indian Patent Act of 1971, moreover, allowed only 'processes' to be patented and 'products' could not be patented. This has now been rectified and made TRIPS compatible under the WTO Agreement. The Patents (Amendment) Act, 2002 provides for patent protection to inventions for a period of twenty years from the date of filing.

There is also a need to remove administrative hassles in granting funds to research scholars/institutes. It has been pointed out that wherever a decentralized mode of implementation has been introduced this has yielded better results. It has been observed, for instance, that research grants offered by the UGC is first passed on to the Registrar of the University who forwards it to the Supervisor/College Principal *for finally releasing the money to the research scholar*. The subsequent release of grant is, similarly, based on availability of 'utilization certificate'. All this consumes a great deal of time. The administrative procedure could be simplified through authorizing the Department or the College Head to receive the grant directly and release the same to the research scholar immediately.

R&D infrastructure with respect to libraries and laboratory facilities also need to be strengthened. Both libraries and laboratory facilities at the college and university levels appear to be treated as adjuncts to these institutions rather than an integral part of the education system. The condition of libraries, at least, might be improved not only through higher government expenditure but a library movement in the country, whereby people may be asked to donate books from their personal collections in return for the assurance of the good upkeep of books. It is worth mentioning that the Constitution Seventy Third Amendment lists libraries as one of the functions to be passed on to the PRIs (Panchayat Raj Institutions). Importance of libraries has been recognized for children as well as it is evident from the pronouncement such as, "no library, no school". New life ought to be instilled in the intellectual arena through all such visions and for laying the foundation of a knowledge based society.

Appendix I

Grants-in-aid of Government of India for Extramural Research

MINISTRY OF HUMAN RESOURCES DEVELOPMENT

(a) Indian Council of Social Sciences Research (ICSSR)

Subject: Social Science/Business Administration

Type of fellowships and eligibility :

(i) National Fellowship: Distinguished Social Scientist

(ii) Sr. Fellowship: Professors, Readers and Sr. Lecturers who have number of publications in their names.

(iii) Gen. Fellowship: Post Doctoral students or young lecturers.

For further information contact:

ICSSR,
Aruna Asaf Ali Marg,
JNU Institutional Area,
New Delhi – 110 067
Email: icssr@ren.nic.in
Website: www.icssr.org

(b) All India Council for Technical Education (AICTE)

Subjects : Mechanical, Products & Industrial Engineering, Chemical Engineering, Civil Engineering, Computer Science, Electronics and Communication, Electrical Engineering, Material Science, Energy Science & Metallurgy, Biotechnology.

Type of grants : Grants for equipment, product, prototype development, expansion of facilities, industrial based R&D output, patent, research, technology know-how, hiring services, technology transfer, writing of technical books, seminars, conferences (not more than Rs.2.5 lakhs).

Eligibility : All IITs, IIMs, University Departments, Deemed Universities, RECs, Government aided engineering colleges and recognised by AICTE.

For further information contact:

Member-Secretary,
Council for Technical Education,

IG Complex, IP Estate,
New Delhi – 110 002
Email: jpg@aicte.emet.in
Website: www.aicte.ernet.in

(c) University Grants Commission (UGC)

(i) Major Research Project
Subject: Any Subject
Type: Research Fellowship
Eligibility: Permanent/Regular working or Retd. Teachers in the universities and colleges (Under Section 2 (F) & 12(B) of UGC Act, 1956)
Duration: Three years.

(ii) Minor Research Project
Subject: Any Subject
Type: Doctorate/Research Fellowship
Eligibility: Permanent/Regular working or Retd. Teachers in the universities and colleges (Under Section 2 (F) & 12(B) of UGC Act, 1956)
Duration: Two years

For further information contact:
University Grants Commission,
South Campus,
Delhi University,
Benito Juarej Marg,
New Delhi – 110 021
Website: www.ugc.org.

(d) Indian Institute of Advanced Study (IIAS)

Subject: Humanities, Indian Culture/Comparative Religion, Social and Natural Sciences.
Type: Fellowship
Eligibility: An eminent scholar
Duration: 3 months to two years

For further information contact:

IIAS,
Rashtrapati Niwas,
Shimla – 171 005, Himachal Pradesh
Website: www.iias.org

(e) Indian Council for Historical Research (ICHR)

Subject: History of Art, Literature, Philosophy, Science and Technology, Epigraphy, Archaeology, Socio-Economic Formations and Allied subjects.

Type: Senior and Junior Research Fellowships

Eligibility: Historians in university, colleges and other institutions. M.Phill.Ph.D students are eligible for Junior Research Fellowships (JRF). M.Phill students are eligible for a grant of fellowship for a period of one year. They can apply to the Council after completing six months of registration.

Duration: 2 to 3 years.

For further information contact:

ICHR,
35, Ferozshah Road,
New Delhi – 110 001

(f) Indian Council of Philosophical Research (ICPR)

Subject: Philosophy & Allied Discipline

Type: Senior Fellowship

Eligibility: Scholars who have made a significant contribution in their areas of research as evidenced from their publications of books, Research papers and articles etc.

For further information contact:

ICPR,
36, Tughlakabad,
M.B. Road, (Near Batra Hospital),
New Delhi – 110 062.
Website: www.icpr.nic.in

MINISTRY OF SCIENCE AND TECHNOLOGY

Council of Scientific and Industrial Research (CSIR)

(i) *Subject:* Science and Technology and Medical Science

Type: Fellowship

Eligibility: M.Sc., B.E. B.Tech. M.E., M.Tech, MBBS, BDS with a minimum of 60% marks. Candidates should not be more than 32 years of age.

Duration: Two years

For further information contact:

Head HRD Group,
CSIR Complex,
Opp. Institute of Hotel Management,
Pusa, New Delhi – 110 012
Website: www.csirhrdg.nic.in

(ii) *Subject*: Science, Engineering, Medicine

Type: Associate ship

Eligibility: People having doctorate (Ph.D, M.D. M.S., MDS) or equivalent degree or having three years of research, teaching or design and development experience are eligible.

Age: Not more than 35 years

Tenure: One year (extended at the discretion of CSIR)

For further information contact:

Head, HRD Group,
CSIR Complex,
Opp. Institute of Hotel Management,
Pusa, New Delhi – 110 012
Website: www.csirhrdg.nic.in

(iii) *Subject:* Science (Chemical, Agricultural, Computer, Medical, Social)

Type: Associateship

Eligibility: Qualified Indian Scientist, Engineers, Technologists, and Medical Personnel returning from foreign countries.

Duration: Three years

For further information contact:

Head, HRD Group,
CSIR Complex,
Opposite Institute of Hotel Management,
Pusa
New Delhi – 110 012.
Website: www.csirhrdg.nic.in

MINISTRY OF HEALTH & FAMILY WELFARE

(a) Indian Council of Medical Research (ICMR)

Subject: Bio-Medical Science

Type: Fellowship

Duration: Three years

Eligibility:

(a) *Senior Research Fellowship* : M.Sc. (Life Science), MA (Social Science), Master in Medical Social Works, Degree holders with 2 years of Research Experience and MBBS/ M.Sc./M.Pharma/ME/M.Tech Degree holders.

(b) *Research Associateship* : MD/MDS/Ph.D in any field

For further information contact:

ICMR,
Ansari Nagar,
P.O. Box 49911,
New Delhi – 110 029.
Website: www.icmr.in

MINISTRY OF AGRICULTURE

Indian Council of Agricultural Research (ICAR)

Subject: Agriculture, Veterinary & Allied Sciences

Type: Fellowship; No. of Fellowships: 475

Eligibility: Candidates having Bachelor's Degree in the field of Agriculture, Veterinary, Agriculture Engineering, Fisheries, Horticulture, Home Science, Sericulture, Dairy Technology Dairy Engineering, Forestry, Agricultural Marketing.

For further information contact:

Controller of Examination,
Education Division; Room No.226,
ICAR, Krishi Anusandhan Bhavan II,
Pusa, New Delhi – 110 012.
Website: www.icar.org.in

MINISTRY OF RURAL DEVELOPMENT

Council for Advancement of People's Action & Rural Technology (CAPART)

Voluntary Organizations (VOs) working in rural areas with legal status as a society registered under Societies Registration Act, 1860 or under the corresponding State Act, as a trust registered under the Indian Trusts Act, 1982 or registered under the Charitable and Religious Act, 1920 are eligible to apply for financial assistance. The VOs/NGOs should further meet the following conditions:

- The Registration of the organization should be three year old.
- The organization should have a proven track record of working with people's participation.
- The organization should have a bank or post office account for the last three years.
- The organization should have Permanent Account No.(PAN) of the Income Tax Department.
- The organization should not have been put under funding restrictions.
- The area of operation of the projects must necessarily be rural.
- The organization should have successfully completed at least one project either through CAPART assistance or some other institutional source of funding.
- The organization should be broad based and representative in character. It should not be a family affair.
- The organization should have good rapport with the people for whom they propose to work and also with panchayat/block/district state development authorities and technical institutions.

For further information contact:

D.G.,
CAPART,
India, Hebitat Centre,
Lodhi Road,
New Delhi – 110003.
Website : www.capart.nic.in

PLANNING COMMISSION

(a) Socio-Economic Research Division

Subject: Social-Economic Research
Type: Grant-in-aid for research studies to: Research Institutions, Individual Research Workers and for organizing Workshops/ Seminars
Eligibility:

(i) *Research Institution (External Proposal)* : a. Research Institutions/Universities whose Annual Report is placed on the Table of the Lok Sabha, b. Research Institutions who have received annual recurring grants from UGC/ICSSR during the last three years, c. Research Institutions which has successfully executed a research project of Rs.5 lakhs or more granted by ICAR, ICMR, CSIR or similar government organization during the preceding five year, d. An institution declared eligible by the Planning Commission based on the assessment of research output, competence of full time faculty and infrastructure available within the institute.

(ii) *Individual Researcher* : a.Researcher should be a scholar of repute.

Period: Period of study should normally not exceed two financial years.

For further information contact:

Adviser,
SER Division,
Planning Commission,.
Yojana Bhavan, Sansad Marg,
New Delhi – 110 001.
Website: www.planningcommission.nic.in

Appendix II

Fellowships available to Indian Students and Scholars in Institutions/Universities Abroad

ALL SUBJECTS

1. Overseas Research Student Award Scheme (ORSAS)

Subjects: All Subjects.
Level of Study: Doctorate and Predoctorate.
Type: Scholarship (for two years with a possibility of renewal for a third year).
Eligibility: Overseas (non-EU) students of outstanding merit having (graduate) degree and having good knowledge of English..
Country of Study: Universities in United Kingdom.

For further information contact:

Higher Education Policy,
Overseas Research Programme,
Universities UK Office,
Weburn House,
20, Tavistock Square,
London, WC1HGHQ,
Website: universities.uk.ac.uk.

2. Cambridge Nehru Scholarships for Ph.D Study.

Subjects: All Subjects
Level of Study: Doctorate, Predoctorate.
Type: Scholarship (for three years)
Eligibility: Applicants must win an Overseas Research Student (ORSAS) award having a First-class Masters' Degree or its equivalent, in addition to a first-class (honours) degree.
Country of Study: United Kingdom. (University of Cambridge)

For further information contact:

The Nehru Trust for Cambridge University,
Teen Murti House, Teen Murti Marg,
New Delhi – 110 011 (India)
Website: www.admn.cam.ac.uk'

3. Dharam Hinduja Cambridge Scholarships

Subjects: All Subjects
Level of Study: Postgraduate

Type: Scholarship

Eligibility: Must win an Overseas Research Student Award (ORSAS). Preference will be given to applicants having a first class Masters Degree or its equivalent in addition to a first class (Honours) degree. Should be less than 35 years of age

Country of Study: United Kingdom University of Cambridge).

For further information contact:

The Nehru Trust for Cambridge University,
Teen Murti House, Teen Murti Marg,
New Delhi – 110 001(India).
Website: www.admn.cam.ac.uk

4. Jawaharlal Nehru Memorial Trust Cambridge Scholarships

Subjects: All Subjects

Level of Study: Doctorate and Predoctorate

Type: Scholarships (two years)

Eligibility: Must win an Overseas Research Student Award (ORSAS). Preference will be given to applicants having a first class Masters Degree or its equivalent in addition to a first Class (Honours) Degree.

Country of Study: United Kingdom (University of Cambridge)

For further information contact:

The Nehru Trust for Cambridge University,
Teen Murti House, Teen Murti Marg,
New Delhi – 110 011 (India).
Website: www.admn.cam.ac.uk

5. Jawaharlal Nehru Memorial Trust Cambridge DFID Scholarships

Subjects: All Subjects.

Level of Study: Postgraduate.

Type: Scholarship (one year).

Eligibility: Applicants must be under 35 years of age on October 1st. Priority will be given to those under the age of 30. Must return to their own country after completing the course, not be employed by a government department or by a parastatal organization. Should not be at present be living or studying in a developed country.

Country of Study: United Kingdom (University of Cambridge).

For further information contact:

The Nehru Trust for Cambridge University,
Teen Murti House, Teen Murti Marg,
New Delhi – 110 011 (India).
Website: www.admn.cam.ac.uk

6. Nehru Centenary Chevening Cambridge Scholarship

Subjects: All Subjects
Level of Study: Graduate (administered by British Council)
Type: Scholarship (2 years)
Eligibility: Applicants must apply and secure first class or high second class (Honours) degree or equivalent. Should be under the age of 26.
Country of Study: United Kingdom

For further information contact:

The Nehru Trust for Cambridge University,
Teen Murti House, Teen Murti Marg,
New Delhi – 110 011, India.
Website: www.admn. cam.ac.uk

7. Michael Miliffe Cambridge Scholarship

Subject: All Subjects
Level of Study: Graduate, Post Graduate
Type: Scholarship (2 years)
Eligibility: Applicants must apply to the University of Cambridge in the regular way and must have a first class/high second class (Honours) degree or equivalent. Must be under 26 years.
Country of Study: United Kingdom (University of Cambridge)

For further information contact:

Website: www.admin.cam.ac.uk/univ.

8. Robert S McNamara Fellowships at Cambridge

Subjects: All Subjects
Level of Study: Postgraduate
Type: Scholarship
Eligibility: Open to students with a Graduate degree.
Country of Study: United Kingdom (University of Cambridge)

For further information contact:

Website: www.admin.cam.ac.uk/univ

9. Radhakrishna/British Chevening Scholarships

Subjects: All Subjects
Level of Study: Postgraduate for one year
Type: Scholarship
Eligibility: Open to Indian nationals (administered by British Council)
Country of Study: United Kingdom

For further information contact:

International Office,
University Office,
Wellington Square, Oxford,
Oxfordshire, OX1, 2JD, England.

10. Rhodes Scholarships

Subjects: All Subjects
Level of Study: Graduate, Postgraduate/Doctorate.
Type: Scholarship (2-3 years)
Eligibility: Candidates must have graduated from a university and have resided in their home country for a number of years. Should be less than 25 years in age
Country of Study: United Kingdom (University of Oxford)

For further information contact:

International Center for Genetic Engineering and Biotechnology,
Aruna Asaf Ali Marg,

New Delhi – 110 067.
India.
Website: www.rhodesscholar.org

11. Association of Rhodes Scholars in Australia

Subjects: All Subjects
Level of Study: Postgraduate Research.
Type: Grant to a maximum, of $20,000
Eligibility: Open to men and women currently engaged in higher research of a Commonwealth University, approved by the committee administering the scholarship. Must be a commonwealth citizen, and should not be graduate of an Australian or New Zealand University.
Country of Study: Australia

For further information contact:

Assistant Secretary,
Association of Rhodes Scholars in Australia,
Travel Scholarship Committee,
Melbourne Research and Innovation Office,
Level 5, 161 Bagy Street,
The University of Melbourne,
Victoria 3010,
Australia
Website: www.unimelb.edu.in

12. Mountbatten Memorial Grants to Commonwealth Students

Subjects: All Subjects
Level of Study: Postgraduate
Type: Grant-in-aid
Eligibility: Students pursuing a course of study at Postgraduate level at common wealth universities or other recognized institutes of higher education in the UK.
Country of Study: United Kingdom

For further information contact:

199, West Malvern Road,

Worcestershire,
WR 14 4BB, England,
Email: casseltrust@sherborn.demon.co.uk
Website: www.britac.ac.uk/funding/guide.

13. New Zealand Commonwealth Scholarships

Subject: Study undertaken should be of developmental relevance to the candidates' home country.
Level of Study: Postgraduates
Type: Scholarships (up to 3 years)
Eligibility: Open to graduates who have graduated within the last 5 years.
Country of Study: New Zealand

For further information contact:
Scholarship Officer,
P.O. Box 11-915 Wellington, New Zealand
Email: schools@nzvcc.ac.nz
Website: www. pzvcc.ac.nz.

14. Huygens Scholarships Programme

Subjects: All Subjects.
Level of Study: Post doctorate or postgraduate.
Type: Scholarship (3-10 months)
Eligibility: Must intend to pursue a specialization at a higher educational institution, recognized by the Dutch Government or at an institute carrying out work related to the academic work they are doing in their own country. Provisional admission to a Dutch higher education institution or a research institute is a necessity. Must also be in their final stages of studies. Must be proficient in English or Dutch, and must not be older than 35 years of age.
Country of Study: Netherlands

For further information contact:
The Information Officer,
NUFFIC

P.O. Box 29777, The Hague, NL-2502 LT,
Netherlands.
EMail: nuffic@nuffic.nl.
Website: www.nuffic.nl.

15. Northumbria University International Scholarships

Subjects: All Subjects
Level of Study: Postgraduate Research, MBA
Type: Scholarship (one year minimum)
Eligibility: Open to non European Union Students beginning a full time course of at least one year's duration.
Country of Study: United Kingdom
For further information contact:

International Office,
Ellison Place Newcastle upon tyse,
NE1 8ST, England.
Email: er.scholarships@northumbria.ac.uk
Website: www.northumbria.ac.uk/nbs

16. One World Scholarship Programme

Subjects: All Subjects
Level of Study: Unrestricted.
Type: Scholarship
Eligibility:18-35 years of age. Preference will be given to candidates who are able to speak German and must have already started their studies in Australia.
Country of Study: Austria
For further information contact:

The Study Adviser,
Student Division,
Turkenstrane 3, Vienna, A-1090,
Austria.

17. AAUW International Fellowship Program

Subjects: All Subjects

Level of Study: Master's/Doctorate/Post doctorate/Professional

Type: Fellowship (One year)

Eligibility: Women with A United States Bachelor's degree or equivalent and proficiency in English language.

Country of Study: USA

For further information contact:

The Administrative Officer,
American Association of University Women,
111116 Street North West, Washington, D.C., 20036, USA.
Email: foundation@ aauw.org.
Website: /www.aauw.org.

18. PEO International Peace Scholarship

Subjects: All Subjects

Level of Study: Doctorate, Postgraduate.

Type: Grant- in- aid (maximum two years)

Eligibility: Based on financial need, nationality, degree, full-time status and residence for the entire term of study.

Country of Study: USA or Canada

For further information contact:

The Project Supervisor,
3700 Grand Avenue, Des Moines, 1A,
www: http://peointernational.org.

19. Catholic University of Louvain Co-Operation Fellowships.

Subjects: Subjects relevant to third –world development.

Level of Study: Doctorate, Postgraduate.

Type: Fellowship (up to four years).

Eligibility: Candidates who have the qualifications required to be admitted at the postgraduate level at UCL. Excellent academic background, with some professional experience and good command of French (DELF). Must be less than 35 years of age for Ph.D. or 40 years of age for a Specialization.

Country of Study: Belgium

For further information contact:

The Catholic University of Louvain (UCL)
Secretariat a la cooperation Internationale Halles Universities,
Place de l' universite 1, Louvain-la-Neuve,
B-1348, Belgium,
Email: baeyens @ sco.ucl.ac.be
Website: www.ucl.ac.be

20. J.N. Tata Endowment for Higher Education

Subjects: All Subjects.
Level of study: Doctorate post doctorate, Postgraduates.
Type: Loan/Gift scholarship (125)
Eligibility: Graduates with good academic and extra curricular record. Deserving mid-carrier professionals are also eligible.
Country of study: Any country except India.

For further information contact:

Administrative Office,
J.N. Tata Endowment,
Mulla House, 4th Floor,
51, M.G. Road, Bombay – 400 001
India.
Website: www.tata.com/tata.sons

21. IES Tokyo Scholarships

Subjects: All Subjects
Level of Study: Postgraduate
Type: Scholarship
Eligibility: Students belonging to the consortium of select colleges and universities belong to the Institute of International Education of Students.
Country of Study: Consortium Countries

For further information contact:

1-21-6 Dogenzaka Shibuya Kku,

Tokyo 150 Japan
Email: Tqzaiden @246.ne.JP
Website: ies.abroad.org

22. Commonwealth Scholarship and Fellowship Plan (CSFP)

Subjects: All Subjects
Level of Study: Postgraduate/Doctorate
Type: Scholarship/Fellowship
Eligibility: Open to graduates and post graduates of Commonwealth Countries.
Country of Study: Commonwealth Countries offering the fellowship during the year (e.g. Australia, Canada, South Africa, Mauritius, Malaysia, New Zealand etc.)

For further information contact:

Indian Council of Cultural Relations (ICCR).
Azad Bhavn,
Indraprastha Estate,
New Delhi – 110002.
Email: iccr@giasdl01.vsnl.net.in
Website: www.sholarshipinindia.com

SOCIAL AND BEHAVIOURAL SCIENCES

1. Robert S. Mcnamara Fellowship Programme (RSM)

Subject: Development Economics
Level of Study: Post-graduate research/full time study.
Type: Fellowship-sponsored by World Bank and Princeton University.
Eligibility: Scholars belonging to Universities, all over the world.
Country of Study: Any University/institute of repute in any country.

For further information contact:

The Scholarship Administrator,
World Bank,
1818 H Street West,
Washington, DC,
U.S.A.
Email: jjwbgsp@worldbank.org.
Website: www.worldbank.org/wbc/scholarships

2. Margaret McNamara Memorial Fund

Subject: Social sciences, social work, public health, public administration, agriculture etc.
Level of Study: Postgraduate
Type: Grant
Eligibility: Open to women who are nationals of developing countries, residing in United States but not as permanent residents. They must be at least 25 years old.
Country of Study: USA

For further information contact:

Chairman of the Selection Committee.
1818, H Street Northwest,
Room Q5-080, Washington, D.C.
20433, USA
Email: wservices1@worldbank.org

3. The Joint Japan/World Bank Graduate Scholarships Programme (JJ/WBGSP)

Subject: Development related fields.
Level of Study: Post-graduate studies.
Type: More than 2000 scholarships.
Eligibility: Scholars belonging to universities, all over the world and belonging to member-countries of the world bank.
Country of Study: All universities/institutes renowned for their development research and teaching.

For further information contact:

The Scholarship Administrator,
World Bank,
1818 H Street West,
Washington, DC,
Email: jjwbgsp@worldbank.org.
Website: www.worldbank.org/wbc/scholarships

4. Ford Foundation International Fellowship Programme

Subject: Applied Social Sciences, environments, arts & culture.
Level of Study: Flexible, as part of the placement process (not exceeding three years).
Type: Fellowship (Full or Partial)
Eligibility: Indian citizen currently residing in Bihar, Jharkhand, Rajasthan, Uttar Pradesh or Uttranchal. (a) should have bachelor's or master's degree with 3 years full-time work experience relevant to the study (b) candidates applying for Ph.D should have Master's degree with 60% marks. For both (a) and (b) proficiency in language of the host institution is a must.
Country of Study: Any Country.

For further information contact:

International Fellowship Process,
Ford Foundaztion,
12, Hailey Road,
New Delhi – 110 001
Email: ask@ifpsa.org.
Website: www.fordfound.org.

5. Cambridge Overseas Studentship in Social Anthropology (ORS equivalent)

Subjects: Social Anthropology
Level of Study: Doctorate
Type: Studentship (2 years)
Eligibility: All overseas students
Country of Study: United Kingdom

For further information contact:

Admission Secretary,
Department of Social Anthropology,
Frer School Lane,
Cambridge, Cambridgeshire,
CB2 3RF, England (United Kingdom).

6. EF Behavioural Science Research Fellowship

Subjects: Epilepsy research relative to the behavioural sciences and other fields of study relevant to epilepsy research and practice.
Eligibility: Open to applicants who have received their doctoral degree in a field of the behavioural sciences by the time the Fellowship commences. Applications from women and minorities are encouraged.

For further information contact:

Epilepsy Foundation,
4351, Garden City drive, Landover, M.D.
20785, USA
www.: http://www.efa.or

7. Human Studies Fellowships

Subjects: Social & Behavioural Sciénce etc.
Level of Study: Graduate, Postgraduate
Type: Fellowship
Eligibility: Open to graduate students with junior or senior standing in the next academic year at accredited colleges and universities.
Country of Study: Any Country.

For further information contact:

The student Coordinator
Institute for Human Studies (IHS)
3301 North Faifax Drive,
Suite 440, Arlington, VA, 22201-4432, USA
Email: his@gmu.edu.
Website: www.http://theihs.org

8. USEFI Post-doctoral Visiting Lecturer Grants

Subjects: Contemporary issues significant to India and the US on the fields of humanities or the social sciences.
Level of Study: Visiting Lecturer Grants (8-10).
Eligibility: Resident Indian citizens who have not visited the USA for the preceding three years. Proficiency in English. Must be full time faculty members of an Indian university or research institute, hold a Ph.D degree or have equivalent published work, with at least 10 years of teaching experience in a college or university. Must not be more than 50 years of age.
Country of Study: U.S.A.

For further information contact:

The United States Educational Foundation in India,
"Fulbright House",
New Delhi- 110 001
Email: info@fulbright-India.Org.

RELGION AND THEOLOGY

1. Harvard University, Centre for the Study of World Religions, Senior Fellowships

Subjects: Religious studies including the historical and comparative study of religions.
Level of Study: Post doctoral
Eligibility: Open to post doctoral scholars
Type: Fellowship
Country of Study: USA

For further information contact:

Coordinator of Education Planning
42 Frances Avenue, Cambridge, MA 02138
USA
Email: brookepalmir@harvard.edu
Website: www.hds.harvard.edu/c&wr

EDUCATION

1. Halen M. Robinson Award

Subject: Literary Education
Level of Study: Doctorate
Type: Reading or literacy Research Grant
Eligibility: Open to all doctoral students worldwide who are members of the International Reading Association
Country of Study: Any Country.

For further information contact:

Email: research@reading.org
www.: http://www.reading.org

2. Reading/Literacy Research Fellowship

Subjects: Literacy education and Education Research
Level of Study: Post doctorate
Type: Fellowship
Eligibility: Open to any International Reading Association member outside the USA or Canada
Country of Study: Any Country.

For further information contact:

Email: research@reading.org
www.: http://www.reading.org

3. CIMO Fellowships for young Researchers and University Teaching Staff

Subjects: Education and research in all subjects
Eligibility: applicants should be under 35
Level of study: Doctorate, Post doctorate, PG Pre-doctorate Research
Type: Scholarship
Country of Study: Finland

For further information contact:

Grants Management Officer
Centre for International Mobility (CIMO)

P.O. Box 343
Hakaniemenkatu 2, Helsinki,
Fin-00531, Finland.
Email: cimoinfio@cimo.fi

MASS COMMUNICATION

1. Thomson Foundation Scholarship

Subjects: Journalism, radio or TV broadcasting, internet publishing and photojournalism

Level of Study: Professional development

Type: Scholarship

Eligibility: Open to professional journalists and broadcasters with at least three years of experience.

Country of Study: United Kingdom

For further information contact:

The Thomas Foundation,
37, Park Place, Cardiff, CF 103AB, Wales,
United Kingdom.
Email: enquiries@thomfound.co.ok
www: http://www.thoimsonfoundation.co.uk

2. Centre for Advanced Study in the Behavioural Sciences

Subjects: Behavioural sciences, biological sciences and humanities.

Level of Study: Post doctorate

Type: Fellowship (9-12 months)

Eligibility: Applicants must hold a Ph.D

Country of Study: USA

For further information contact:

The Associate Director
75 Alla Road, Stanfond, CA 94305, USA

3. Computer Laboratory QRS Equivalent Awards

Subjects: Computer Science

Level of Study: Postgraduate

Type: Award (upto 3 years)

Eligibility: Candidates studying for a Ph.D in the Computer Lab.

Country of Study: United Kingdom

For further information contact:

Website: www.admin.cam.ac.uk/univ.

3. The Countess of Munster Musical Trust Awards

Subjects: Musical Studies
Level of Study: Doctorate, PG, Professional development
Type: Grant (1 year may be renewed)
Eligibility: British Commonwealth citizens. 18-24 years of age for instrumentalists and composers or under 28 for singers
Country of Study: Any Country.

For further information contact:

Secretary
Wonmley Hill, Godalming, Sunrey, GU8 5SG, England
Email: munstertrust@compuserve.com
Website: www.munstertrust.ukgrateway.net.

4. Royal Overseas League Annual Music Competition

Subjects: Musical Performance –Instrumental and Vocal.
Level of Study: Professional development
Type: Competition
Eligibility: Not above 28 years of age for instrumentalists and 30 years of age for singers.
Country of Study: Any Country.

For further information contact:

The Promoter Officer
Royal Overseas League,
Over-seas House, Park Place,
St.James Street, Londfon, SWIA ILR, England
Email: Culture@rosl.org.ok
Website: www.rosi.org.uk.

5. Eastman School of Music Graduate Awards

Subject: Music
Level of Study: Doctorate, PG
Type: Tuition Scholarship & Stipends (1 year, renewable)
Eligibility: Open to nations of all countries. Should have the qualifications necessary for admission to the Eastmas School of Music.

Country of Study: USA

For further information contact:

The Director - Financial Aid.
Eastman School of Music of the University of Rochester
26, Gibbs Street, Rochester, NY 14604, USA
Email: esmadmit@uhura.cc.rochester.edu.
Website: www.rochester.edu/eastman

ARCHITECTURE

1. Institute for Housing & Urban Development Studies Fellowship for Courses

Subjects: Housing planning & building
Level of Study: Postgraduate
Type: Fellowship
Eligibility: Applicants from any country with relevant professional experience and engaged in work related to urban management and housing. Proficiency in English.
Country of Study: Netherlands

For further information contact:

Head PC&M Bureau
P.O. Box 1935, Rotterdam, NL-3000BX,
Netherlands
Email: ibs@ibs.nl
Website: www.ilb.nl

AGRICULTURE, FORESTRY AND FISHERY

1. International Foundation For Science (IFS) Grant

Subject: Research related to projects in agriculture, forestry, natural products and aquatic resources and national ecosystems.
Type: Research Grant
Eligibility: Applicants must have a M.Sc. degree or its equivalent and must be attached to a university/research institution in a developing country and should be below 40 years of age.
Country of Study: Developing Countries

For further information contact:

Head of Finance and Administration
I.F.S.,
19, Stockholm,
S-11438, Sweden.
Email: info@ifs.se
www.: http://www.ifs.se

2. TWAS Grants for Scientific Meetings in Developing Countries

Subjects: Agriculture and other sciences
Level of Study: Postgraduate, Professional Development.
Type: Travel Grant
Eligibility: Open to organizers of international scientific meetings in developing countries.
Country of study: Developing Countries.

For further information contact:

Third World Academy of Sciences (TWAS)
Abdus Salem International Centre for Theoratical Physics,
Strada Costiera 11,
34014, Trieste,
Itlay
Email: info@twas.org.
Website: www.Twas.org.

3. TWAS South _South Fellowship Scheme

Subjects: All fields of Basic Sciences
Level of Study: Postgraduate
Type: Fellowship (1-3 months)
Eligibility: Permanent positions in universities/research institutions in developing countries.
Country of Study: Developing Countries
For further information contact:

Third World Academy of Sciences (TWAS)
Abdus Salem International Centre for Theoratical Physics,
Strada Costiera 11,
34014, Trieste,
Itlay
Email: info@twas.org.
Website: www: http://twas.org.

4. TWNSO Grants for Joint Research Projects

Subjects: Major Third World problems related to agriculture and technology.
Level of Study: Joint Research
Eligibility: Open to individuals or institutions credited with scientific or technical innovations.
Country of Study: Any country.
For further information contact:

The Third World Network of Scientific Organization,
TWNSO Secretariat,
ICTP Adriotico Guest House,
7th Floor, Stradacosteria 11,
34014 Trieste, Italy.
Email: info@twnso.org.
Website: www.twnso.org.

5. The Bentley Fellowship

Subjects: Use of forage crops, agricultural production and soil fertility.

Level of Study: Doctorate, Graduate, Postdoctorate, Postgraduate.
Type: Fellowship (4-12 months)
Country of Study: Canada or Developing Countries

For further information contact:

International Development Research Centre,
250, Albert Street, Ottawa,
Ontario, K1 P6 M1,
P.O. Box 8500, Ottawa, Ontario,
K1G3H9,
Website: www.idrc.ca

6. Austrian Academy of Sciences PG Course in Limonology

Subjects: Limnology & related topics.
Level of Study: Postgraduate
Type: Scholarship (6 months)
Eligibility: Candidates from developing countries. Below 25-35 years of age, working knowledge of English with an academic degree in science, agriculture or veterinary medicines from a university or other recognized institutes of higher education. Practical experience of at least one special subject required.
Country of Study: Austria or Czech republic.

For further information contact:

Austrian Academy of Sciences,
Dr. Ignez Seipetpletz 2
A-1010, Viena, Austria.

7. PG Studies in physical land resources Scholarships

Subjects: Agriculture and related topics
Level of Study: Postgraduate
Type: Scholarship (one year within maximum extension of 2 years)
Eligibility: Candidates from the developing countries or non European Union members.
Country of Study: Belgium

For further information contact:

Geological Institute,
Krijgslaan 281/58, Ghent, B-9000,
Belgium
Email: pirprog.adm@rug.ac.be
Website: www.allserv.rug.ac.be

8. Third World Network of Scientific Organizations (TWNSO) Grants to Institutions in the South for Joint Research Projects.

Subject: Agriculture, Limnology & related topics
Level of Study: Professional Development
Type: Grant (up to 2 years)
Eligibility: Candidates from developing countries in the South
Country of Study: Developing Countries

For further information contact:

TWNSO,
C/o TWAS & the Abdus Salam ICTP,
Trieste, 1-34014, Italy.
Email: info@twnso.org.
Website: www.twnso.org.

BUSINESS MANAGEMENT/ENVIRONMENTAL SCIENCE

1. Fullbright CII Fellowships for Leadership in Management

Subject: Leadership in Management
Level of Study: Professional Development
Type: Fellowship
Eligibility: Open to Indian Business Managers
Country of Study: USA

For further information contact:

Deputy Director,
Confederation of Indian Industry (CII),
23, Institutional Area, Lodhi Road,
New Delhi – 110 003
Email: kavita.kohli@ciionline.org.

2. IATF –Aviation MBA Scholarship

Subjects: Aviation Law, business administration and management
Level of Study: PG MBA
Type: Scholarship (one year)
Eligibility: Open to staff from IATA, member airlines from developing countries. Acceptance to the course by Concordia University is required.
Country of Study: Canada

For further information contact:

Concordia University,
Annex FB 801,
1455, &E maisonneuve Boulevearde West,
Mopntreal, H #G IM8,
Canada
Email: iatj@iata.org

3. IATF IATA Aviation Training and Development Institute (AIDI) Scholarships

Subjects: Business Administration and Management in the field of Aviation Training and Development

Level of Study: Postgraduate

Type: Scholarship (3-30 days per course)

Eligibility: Open to staff from IATA member Airlines from countries with developing economies.

Country of Study: Switzerland, USA or Singapore,

For further information contact:

IATA Centre,
Raute de 1 Aeroport 33,
PO Box 416 Geneva, CH-1215,
Switzerland
Email: iatf@iata.org.

4. International Postgraduate Research Fellowship (IPRS)

Subjects: Health science, humanities, social sciences, science, technology, engineering, law or management

Level of study: Doctorate, Postgraduate

Type: Scholarship (two years for the Master's Degree and three years for the Ph.D.)

Eligibility: Candidates will be selected on the basis of academic merit and research capacity. Open to qualified overseas graduates.

Country of study: Australia

For further information contact:

Email: international@latrobe.ed.au

5. Shell Centenary Chevening Scholarship for Postgraduate Study

Subjects: Applied Sciences and technology,including environmental sciences, business management or economics.

Level of Study: Postgraduate

Type: Scholarship (one year)

Eligibility: Applicants must apply to the university of Camdbridge in the normal way/. They must be under the age of 26 and have a first class or high second class (Honours) Degree of equivalent.

Country of Study: United Kingdom (University of Cambridge)

For further information contact:

http://www.admin.cam.ac.uk/univ/gsprospectus/c&/overseas/schemes.html.

6. Golda Meyer Mount Carmel International Training Center Assistance.

Subjects: Courses in community organization and management of human services, pre-school education organization and management of income generating projects and small scale industries and management of non-governmental organizations (NGOs)

Level of study: Postgraduate. Professional development for a period of four and eight weeks.

Eligibility: Age 24-45 years, priority given to women applicants.

Country of study: Israel

For further information contact:

The Director,
Golda Meyer Count Carmel International Training Centre,
P.O. Box 6111,
Haefa, 31060, Israel
Email: mctc@mctc.co.il

7. Golda Meyer Count Carmel International Training Center Tuition and Maintenance Scholarships.

Subjects: Community organizations and management of human services pre-school education, organization and management of income generating projects and small-scale industries or management of non-government organizations (NGOs).

Level of Study: four to eight weeks

Eligibility: Open to nationals of developing countries and transitional societies.

Country of Study: Israel

For further information contact:

The Director
Golda Meyer Count Carmel International Training Centre,
P.O. Box 6111,

Haefa, 31060, Israel
Email: mctc@mctc.co.il

8. British Chevening and London Business School Scholarship

Subject: MBA
Level of study: MBA
Type: Scholarship
Eligibility: Applicants who are short listed by the London Business School will be interviewed by the British Council in India and may be required to travel.
Country of study: United Kingdom.
For further information contact:

MBA Admissions Officer
Regents Park, London NW1, USA, England
Email: mferreira@lon.ac.uk
Website: www.ibs.ac.uk.

9. Insead L'Greal Scholarship

Subject: MBA
Level of study: MBA
Type: Scholarship (one year)
Eligibility: Candidates of any nationality who demonstrate a capacity for creativity, innovation and entrepreneurial activity.
Country of study: France or Singapore
For further information contact:

http://www.insead.ed./mba

10. Insead Misys Foundation Scholarship

Subject: MBA
Level of study: MBA
Type: Scholarship
Eligibility: Open to candidates from developing countries admitted to the MBA programme.
Country of Study: France or Singapore

For further information contact:

Website : www.insead.ed./mba

11. INSEAD Sasakawa (Syllf) Scholarships

Subjects: MBA

Level of study: MBA

Type: Scholarship

Eligibility: Candidate of any nationality are eligible

Country of Study: France or Singapore

For further information contact:

http://www.insead.ed./mba

12. Chevening Scholarships

Subjects: MBA

Level of study: MBA

Type: Scholarship

Eligibility: Applicant must be residents in their home country at the time of applying -

Country of study: England—Applications must first be submitted to the British Council.

For further information contact:

Email: m.Williams@cransfield.ac.uk

Website: www.cranfield.ac.uk/som.

MEDICAL SCIENCES

1. International Fellowship in Medical Education

Subjects: Eligible areas of study include educational methodology, curriculum design, evaluation systems, medical school governance and the development of basic and clinical science departments

Level of Study: Professional development

Type: Fellowship (6-12 months)

Eligibility: Applicants must have a graduate or professional degree in clinical medicine, a basic science or education and must have a good knowledge of English. More than three years of experience as a faculty member as well as an independent of a home country medical school is also required.

For further information contact:

The Director, Programme Planning and Development.
3624 Market Stret, Philadelphia, PA,
19104 –2685 USA
Website: www: http://www.ecfmg.org

2. ISN Fellowships Awards

Subject: Nephrology

Level of study: Post-doctorate, Pre-doctorate.

Type: Fellowship (3 months for short term and 1-2 years for long term.)

Eligibility: Open to young Nephrologists from emerging countries. Sufficient training in internal medicines or other fields. Fellowship offered in clinical training but research training may be allowed in some cases.

For further information contact:

The Administrative Assistant
International Society of Nephrology (ISN)
Academic Medical Centre,
Department of Pathology,
Meibergdreef 9, Amsterdam, NL –11'05 AZ,
Netherlands
Email: isn@amc.uva.nl

3. TWAS Awards in Basic Sciences

Subjects: Basic medical sciences, biology, chemistry, mathematics and physics.
Level of study: Doctorate, Ph.D, PG and professional development.
Eligibility: Open to nationals of developing countries. Fellows of the Academy of Science are not eligible.
Country of Study: Developing Countries

For further information contact:
Email: info@twas.org
Website : www.twas.org.

4. TWAS Prizes to young scientists in Developing Countries

Subjects: Biology, Chemistry, Mathematics or Physics related annually.
Level of Study: Postgraduate
Type: Prize/Award
Eligibility: Open to academics and research councils in developing countries. Age limit is 40 years
Country of Study: Developing Countries

For further information contact:
Email: info@twas.org
Website: www.twas.org.

5. TWAS Research Grants

Subjects: Biology, Chemistry, Mathematics or Physics
Level of Study: Doctorate, Post-doctorate, Postgraduate, Professional Development.
Eligibility: Nationals of developing countries with an advanced academic degree, some research experience and hold positions at universities or research institutions in developing countries.
Country of Study: Developing Countries

For further information contact:
Email: info@twas.org
Website: www.twas.org.

NATURAL SCIENCE

1. Colt Foundation Ph.D Fellowship

Subjects: Medical and Natural sciences and related topics
Level of Study: Doctorate (3 years)
Eligibility: Open to any student proposing to take a Ph.D. in the above mentioned subjects.

For further information contact:

The Director,
Colt Foundation,
New Lane, Havart, Hompshire, PO9 2 LY,
England
Email: jackee.Douglas@coltgroup.com
Website: coltfoundation.org

2. Jawaharlal Nehru Memorial Fund Cambridge Scholarship for Ph.D Study.

Subjects: Environmental Ecology and sustainable development and other subjects.
Level of study: Doctorate, Post Doctorate
Type: Scholarship
Eligibility: Indian applicants must win an overseas student (ORS) award. Those having a Master's Degree may be given preference.
Country of Study: United Kingdom

For further information contact:

Asstt. Tutorial Secretary,
Merton College, Oxford,
Oxfordshire,
OX1 4JD, England

3. Post graduate studies in Physical land resources scholarship

Subjects: Physical land resources, soil sciences, microbiology, archaeopedology
Level of Study: PG

Type: Scholarship (1 year with maximum extension to 2 years)
Eligibility: Candidates who are nationals of developing countries or non-European Union members.
Country of Study: Belgium

For further information contact:

Geological Institute,
University Grants,
Krijgslaan, 281/S8, Ghent,
B-9000 – Belgium
Email: pisprog.adm@rug.ac.be
Web: www.allsen.rug.ac.be/amtangbe/main.litme.

PS. : For numerous other scholarships offered by colleges, universities and other institutions (list too large to be included here), students and scholars are advised to check on the internet.

Appendix III

Questionnaire

Name : ....

Designation ....

1) Are the Researchers in your Department/Faculty/Institution aware of the grant-in-aid for research by the following institutions? (Please give your reply in Yes/No in the bracket given against each Institution):
   a) UGC
   b) ICSSR
   c) IAAS
   d) ICHR
   e) CSIR
   f) ICPR
   g) ICAR
   h) Planning Commission
   i) Ministry of Environment and Forests
   j) Ministry of Rural Development
   k) Foreign Funding Agencies
2) What are the main causes of low level of awareness in the above context?
   a) Lack of interest motivation ....

   ....

   ....

   b) Lack of effective publicity ....

   ....

   ....

   c) Other causes ....

   ....

   ....

3) What measures should be taken to raise the level of the aforesaid awareness?

....

....

4) How many of the (grant) facilities for research provided by the Institutions noted above have been availed by you/ your Department/Faculty/Institution during the period 1997-98 to 2002-03?

....

....

....

5) State, on the basis of experience, the difficulties in availing the (grant) facilities for research.

a) Cumbersome procedure

....

....

b) Inadequacy of the amount granted

....

....

c) Other difficulties

....

....

6) Have the funded Research Project been completed in time?

....

....

7) In case of delayed competition of the funded Research Project, give the major reasons for the delay.

....

....

....

8) State whether the products of Research work have been published/ reviewed/utilized for further research/used for policy formulation in the related field.

....

....

....

9) In view of the competition to win the grant from the common pool, how does your state fare this competition?

....

....

....

10) Do you commend the policy of the State Government in regard to provision of funds to research institutions?

....

....

....

11) What is your overall opinion in regard to demand for and funding of research projects in the State ?

....

....

....

12) What are your suggestions for improvement in funding of the research projects in the State ?

....

....

....

13) What the areas/issues relating to promotion and funding of research, which are getting neglected at the national level?

....

....

....

14) What are the remedies for the aforesaid neglect?

....

....

....

[Signature]

N.B. : In case the provided space is insufficient, please use additional sheets.

FINE ARTS

1. USEFI Post Doctorate Travel-only Grants

Subjects: Fine Arts or the performing arts, social sciences and humanities

Level of Study: Post doctorate, Professional Development (2)

Type: Grant-in –Aid.

Eligibility: Resident Indian citizens who have not visited the USA for the preceding three years. Proficiency in English. Must be full time faculty members of an Indian university or research institute, hold a Ph.D degree or have equivalent published work, with at least 10 years of teaching experience in a college or university. Must not be more than 50 ytears of age.

Country of Study: USA

For further information contact:

The United States Educational Foundation in India,
"Fulbright House",
New Delhi- 110 001
Email: info@fulbright-India.Org.

2. USEFI Professional Fellowships on Plastics and Performing Arts, Museum Studies and Arts/Culture Management

Subjects: Museum studies, arts or culture management, plastics or performing arts.

Level of Study: Post Graduate; Professional Development

Type: Fellowship (2-6 months)

Eligibility: Resident Indian citizens who have not visited the USA for the preceding three years. High level of academic achievements and a good knowledge of English is required. Age between 25-40 years, having a Post Graduate degree on the relevant area.

Country of Study: USA

For further information contact:

The United States Educational Foundation in India,
"Fulbright House",
New Delhi- 110 001
Email: info@fulbright-India.Org.

Bibliography

Agarwal, S.P., 2002, *Technology Exports and International Competitiveness, Technology Exports, Vol.V (No.1),* Indian Institute of Foreign Tradee, New Delhi.

ACU, 2003, International Awards, 36 Gardon Square (WC1HOPI), London.

Basham, A.L., 2003, *The Wonder That Was India*, Rupa & Co., New Delhi,

Bhatnagar, S.S., 1949, *Inaugural Address, Annual Session of the National Institute of Sciences*, Patna,

Bhattacharya, A.B., 1994, Intellectual Property Policy Proposal for Indian Academic Institute: Intellectual Property Rights, Nair, K.R.G. Allied Publishers Limited, New Delhi.

Bose, P.K., 1995, *Research Methodology*, Indian Council of Social Science Research, New Delhi,

Bowender, B., 2003, R&D in India, *ASCI Issue Paper 9*, Hyderabad: ASCI.

Directory of Grant Making Trusts,2003, Charities Aid Foundation, London.

Chiang, A.C., 1974, *Fundamental Methods of Mathematical Economics*, McGraw Hill Book Company, New Delhi.

Grants Register, 2003, Palgrave, McMillan, London.

Government of India, 1951, Planning Commission, *The First Five-Year Plan: A Draft Outline, New* Delhi: Government of India.

———, 1976, Ministry of Agriculture, *Research and Extensions*, Report of the National Commission on Agriculture (1976), New Delhi.

———, 1986, Ministry of Human Resource Development, *National Policy on Education (1986),* New Delhi

———, 1992, Ministry of Human Resource Development, *National Policy of Education, Programme of Action (1992),* New Delhi.

———, 1993, Department of Education, Ministry of Human Resource Development, *Education for All: The Indian Scene*, New Delhi.

———, 1997, Planning Commission, Ninth *Five Year Plan*, New Delhi: Government of India,

———, 1999-2003, Ministry of Human Resource Development, *Annual Report (1998-99),* New Delhi.

———, 1998-2004, Ministry of Finance, Detailed Demand for Grants for Ministry of Human Resource Development, New Delhi.

———, 1998-2004, Ministry of Finance, Detailed Demand for Grants for Ministry of Science and Technology, New Delhi.

———, 1998-2004, Ministry of Finance, Detailed Demand for Grants for Ministry of Health and Family Welfare, New Delhi.

———, 1998-2004, Ministry of Finance, Detailed Demand for Grants for Ministry of Agriculture, New Delhi.

———, 1998-2004, Ministry of Finance, Detailed Demand for Grants for Ministry of Environment and Forest, New Delhi.

———, 1998-2004, Ministry of Finance, Detailed Demand for Grants for Ministry of Rural Development, New Delhi.

———, 1998-2004, Ministry of Finance, Detailed Demand for Grants for Ministry of Planning, New Delhi.

———, 2001, Planning Commission, *Approach Paper to the Tenth Five Year Plan (2002-2007)*, New Delhi.

———, 2001, Planning Commission, *Report of Steering Committee on Secondary, Higher and Technical Education for the Tenth Five Year Plan*, New Delhi.

———, 2001, Planning Commission (SCSTTP) *Report of Steering Committee on Science & Technology for Technology Promotion for the Tenth Five Year Plan,* New Delhi.

———, 2003, *Tenth Five Year Plan 2002-07*, New Delhi.

———, 2002, Ministry of Science & Technology, *Annual Report of Department of Scientific & Industrial Research (2001-02),* New Delhi.

———, 2000, Ministry of Science and Technology, *Research and development in industry: An overview*, New Delhi.

———, 2002, Ministry of Science and Technology, *Research and Development Statistics 2000-01*, New Delhi.

———, 2004, TIFAC (Technology Information, Forecasting and Assessment Council), Website http://www.tifac.org.in/index1.htm .

———, 2004, TDB (Technology Development Board), *Annual Report 1996-97*, http://www.nic.in/tdb/tdbview.htm .

———, 2004, CSIR (Council of Scientific and Industrial Research), website http://www.csir.res.in/ .

Kalam, A.P.J., and Y S. Rajan, 1998, *India 2020: A vision for the New Millennium*, New Delhi: Penguin Books.

Kuchhal, R.D., 1995, *Vision of Science in India: A Historical Perspective*, World Science News Delhi.

Laws, Sophgie et al, 2003, *Research and Development*, Visten Publication, New Delhi.

Mani, S., 2002, *Government, Innovation and Technology Policy: An International Comparative Analysis*, Cheltenham: Edward Elgar.

Mashelkar, R., 1996, *CSIR: 2001- Vision and Strategy*, New Delhi, CSIR.

National Science Foundation, 1995, *Asia's New High-Tech Competitors*, NSF 95-309, Arlington, VA.

Pal, Yash, 1988, Foreword: Report of the Inter Agency Working Group on Development of Information and Libraries Network, University Grants Commission, New Delhi.

Pepper, Karl R, *The Logic of Scientific Discovery*, Whiley Science Edition, New York, 1961.

Prasad K., *Formulation and Testing of Hypothesis: A Guide to Research in Economics* : C.T. Kurian.

Radhakrishnan, S. and C.S Moore, 1957, *Indian Philosophy*, Princeton University Press, London.

Robbins, Lionel, 1955, *An Essay in the Nature and Significance of Economic Science,* McMillan, London.

Russell, Bertrand, 1949, The Scientific Outlook, George Alen & Unwin Ltd, London.

Russell, Bertrand, 1959, *Wisdom of the West*, McDonald & Co. (Publishers) Ltd., London.

Ryon, John, 1994, *Regional Innovation System in Europe: The Cases of Baden-Wurttemberg and Wales (mimio),* University of Manchester, Manchester.

Sjoberg, G. and Roger Nett, *A Methodology for Social Research*, Harper & Row Publishers, New York, 1968.

Study Abroad, 2002, SIT Occassional Paper Series, School for International Training, World Learning (www.sit.edu), Vermont, USA.

Sulaiman, R., 2003, Agricultural Extension: Involvement of Private Sector, Occasional Paper (29) , NABARD, Mumbai

Taher, M., 2001, Libraries in India's National Development Perspective, Concept Publishing Company, New Delhi.

Tibergen, T., 1967, *Economic Policy: Principles and Design,* North-Holland Publishing Company, Amsterda,.

Toyenbee, Arnold, 1884, *Lectures on the Industrial Revolution in England,* http://socserv2.socsci.memaster.ca/.econ/ugem/3113/toynbee/indrev.

University Grants Commission, 1988, *Development of Information and Library Network,* Computers Prints Combine, New Delhi.

UNESCO, 2003, Study Abroad, 7 Place de Fontenoy, 75700 (France), Paris.

Vyas, S.D. and D.K. Singh, 2003, *Knowledge Management Challenges for Library and Information,* Library Herald, Delhi Library Association.

World Bank, 2000, *India: Scientific and Technical Manpower Development in India,* Report No. 20416-IN, Washington DC.

